ENGINEERING STRUCTURES
Geological Foundations

Kruger Brentt
Publishers

ENGINEERING STRUCTURES
Geological Foundations

S.K. Mandal
Dorothy Mason

Kruger Brentt
Publishers
2023

Kruger Brentt Publishers UK. LTD.
Company Number 9728962

Regd. Office: 68 St Margarets Road, Edgware, Middlesex HA8 9UU

© 2023 AUTHORS

ISBN: 978-1-78715-135-2

For information on all our publications visit our website at http://krugerbrentt.com/

PREFACE

Structure Engineering is the most major field in Civil Engineering. In this field engineers design structures, analyze their stability by the use of software's or manual calculations and then construct them. In the previous five decades structural engineering has evolved to be an interesting and important field, It includes the study of dynamics of a structure, its load capacity, design and economy. Due to an increase in the demand of earthquake-proof buildings, the importance of structural engineers has also come to light. Structural engineering is a field of engineering, more appropriately, civil engineering; It deals with the analysis and design of structures. Structural engineering is the design of structural support systems for buildings, bridges, earthworks, and industrial structures. This branch of engineering focuses on supporting a load safely, and relies on mathematical and Physical concepts for the design of the supporting structures. Structural engineers can design machines, cranes, vehicles and even other non-building structures.

Structural engineering is a field of varied and complex tasks. Structural engineers combine simple and basic structural elements to build up complex structural systems, but the main objective is always safety, and to minimize the risk of collapse. Though aesthetics can also be a priority in some cases. Many of the designs for concrete buildings today are specifically meant to be earthquake and hurricane-resistant. Structural geology is the study of the three -dimensional distribution of rock units with respect to their deformational histories. The primary goal of structural geology is to use measurements of present-day rock geometries to uncover information about the history of deformation (strain) in the rocks, and ultimately, to understand the stress field that resulted in the observed strain and geometries.

The present book contains 26-chapters namely Geology, mineralogy, common rock-forming minerals, petrology, igneous rock and their constituent minerals, sedimentary rocks, common metamorphic rocks, structural geology, groundwater movement, stratigraphy, earthquake, structural geology, plate tectonics, ground investigations, soil strength, slope failure and landslides, glacial deposits, coastal processes, geology and civil engineering, floodplain and groundwater, igneous rock, metamorphism and tamorphic rocks, sedimentary rocks and stratigraphy, geological structures: fold and fault, discontinuities and geomorphological processes. Presents the subject of foundation engineering in a logical framework in a natural sequence. It clearly and concisely explains basic principles in the most important areas of foundation engineering. It explains the understanding of the nature and properties of soil and similar natural or artificial systems upon which foundations are built. It explores the latest information on design, analysis and construction of foundations for a wide variety of purposes, including amelioration of deficient foundation soils. This book is solely written to meet the requirements of undergraduate courses in B.E. (Civil Engineering), B.Tech (Civil Engineering), Environmental engineering and the postgraduate courses of various universities and institutions.

We are grateful to all those persons as well as various books, manuals, periodicals, magazines, journals etc. that helped in the preparation of this book. In spite of the best efforts, it is possible that some errors may have occurred into the compilation and editing of the book. Further queries, constructive suggestions and criticisms for the improvement of the book are always welcomed and shall be thankfully acknowledged.

S.K. Mandal

Dorothy Mason

CONTENTS

GEOLOGY 1

Geology is an earth science comprising the study of solid Earth, the rocks of which it is composed, and the processes by which they change. Geology can also refer generally to the study of the solid features of any celestial body (such as the geology of the Moon or Mars).

Geology gives insight into the history of the Earth by providing the primary evidence for plate tectonics, the evolutionary history of life, and past climates. Geology is important for mineral and hydrocarbon exploration and exploitation, evaluating water resources, understanding of natural hazards, the remediation of environmental problems, and for providing insights into past climate change. Geology also plays a role in geotechnical engineering and is a major academic discipline.

GEOLOGIC MATERIALS

The majority of geological data comes from research on solid Earth materials. These typically fall into one of two categories: rock and unconsolidated material.

Rock

There are three major types of rock: igneous, sedimentary, and metamorphic. The rock cycle is an important concept in geology which illustrates the relationships between these three types of rock, and magma. When a rock crystallizes from melt (magma and/or lava), it is an igneous rock. This rock can be weathered and eroded, and then redeposited and lithified into a sedimentary rock, or be turned into a metamorphic rock due to heat and pressure that change the mineral content of the rock which gives it a characteristic fabric. The sedimentary rock can then be subsequently turned into a metamorphic rock due to heat and pressure and is then weathered, eroded, deposited, and lithified, ultimately becoming a sedimentary rock. Sedimentary rock may also be re-eroded and redeposited, and metamorphic rock may also undergo additional metamorphism. All three types of rocks may be re-melted; when this happens, a new magma is formed, from which an igneous rock may once again crystallize.

The majority of research in geology is associated with the study of rock, as rock provides the primary record of the majority of the geologic history of the Earth.

Unconsolidated material

Geologists also study unlithified material, which typically comes from more recent deposits. These materials are superficial deposits which lie above the bedrock. Because of this, the study of such material is often known as Quaternary geology, after the recent Quaternary Period. This includes the study of sediment and soils, including studies in geomorphology, sedimentology, and paleoclimatology.

WHOLE-EARTH STRUCTURE

Plate tectonics

In the 1960s, a series of discoveries, the most important of which was seafloor spreading, showed that the Earth's lithosphere, which includes the crust and rigid uppermost portion of the upper mantle, is separated into a number of tectonic plates that move across the plastically deforming, solid, upper mantle, which is called the asthenosphere. There is an intimate coupling between the movement of the plates on the surface and the convection of the mantle: oceanic plate motions and mantle convection currents always move in the same direction, because the oceanic lithosphere is the rigid upper thermal boundary layer of the convecting mantle. This coupling between rigid plates moving on the surface of the Earth and the convecting mantle is called plate tectonics.

The development of plate tectonics provided a physical basis for many observations of the solid Earth. Long linear regions of geologic features could be explained as plate boundaries. Mid-ocean ridges, high regions on the seafloor where hydrothermal vents and volcanoes exist, were explained as divergent boundaries, where two plates move apart. Arcs of volcanoes and earthquakes were explained as convergent boundaries, where one plate subducts under another. Transform boundaries, such as the San Andreas fault system, resulted in widespread powerful earthquakes. Plate tectonics also provided a mechanism for Alfred Wegener's theory of continental drift, in which the continents move across the surface of the Earth over geologic time. They also provided a driving force for crustal deformation, and a new setting for the observations of structural geology. The power of the theory of plate tectonics lies in its ability to combine all of these observations into a single theory of how the lithosphere moves over the convecting mantle.

Earth structure

Advances in seismology, computer modeling, and mineralogy and crystallography at high temperatures and pressures give insights into the internal composition and structure of the Earth.

Seismologists can use the arrival times of seismic waves in reverse to image the interior of the Earth. Early advances in this field showed the existence of a liquid outer core (where shear waves were not able to propagate) and a dense solid inner core. These advances led to the development of a layered model of the Earth, with a crust and lithosphere on top, the mantle below (separated within itself by seismic discontinuities at 410 and 660 kilometers), and the outer core and inner core below that. More recently, seismologists have been able to create detailed images of wave speeds inside the earth in the same way a doctor images a body in a CT scan. These images have led to a much more detailed view of the interior of the Earth, and have replaced the simplified layered model with a much more dynamic model.

Mineralogists have been able to use the pressure and temperature data from the seismic and modelling studies alongside knowledge of the elemental composition of the Earth to reproduce these conditions in experimental settings and measure changes in crystal structure. These studies explain the chemical changes associated with the major seismic discontinuities in the mantle and show the crystallographic structures expected in the inner core of the Earth.

Geologic time

The geologic time scale encompasses the history of the Earth. It is bracketed at the old end by the dates of the earliest Solar System material at 4. 567 Ga, (gigaannum: billion years ago) and the age of the Earth at 4. 54 Ga at the beginning of the informally recognized Hadean eon. At the young end of the scale, it is bracketed by the present day in the Holocene epoch.

Important milestones:

- 4. 567 Ga: Solar system formation
- 4. 54 Ga: Accretion of Earth
- c. 4 Ga: End of Late Heavy Bombardment, first life
- c. 3. 5 Ga: Start of photosynthesis
- c. 2. 3 Ga: Oxygenated atmosphere, first snowball Earth
- 730–635 Ma (megaannum: million years ago): second snowball Earth
- 542 ± 0. 3 Ma: Cambrian explosion – vast multiplication of hard-bodied life; first abundant fossils; start of the Paleozoic
- c. 380 Ma: First vertebrate land animals
- 250 Ma: Permian-Triassic extinction – 90% of all land animals die; end of Paleozoic and beginning of Mesozoic
- 66 Ma: Cretaceous–Paleogene extinction – Dinosaurs die; end of Mesozoic and beginning of Cenozoic

- ⦿ c. 7 Ma: First hominins appear
- ⦿ 3. 9 Ma: First Australopithecus, direct ancestor to modern Homo sapiens, appear
- ⦿ 200 ka (kiloannum: thousand years ago): First modern Homo sapiens appear in East Africa

BRIEF TIME SCALE

The following four timelinesshow the geologic time scale. The first shows the entire time from the formation of the Earth to the present, but this compresses the most recent eon. Therefore the second scale shows the most recent eon with an expanded scale. The second scale compresses the most recent era, so the most recent era is expanded in the third scale. Since the Quaternary is a very short period with short epochs, it is further expanded in the fourth scale. The second, third, and fourth timelines are therefore each subsections of their preceding timeline as indicated by asterisks. The Holocene (the latest epoch) is too small to be shown clearly on the third timeline on the right, another reason for expanding the fourth scale. The Pleistocene (P) epoch. Q stands for the Quaternary period.

Dating Methods

Geologists use a variety of methods to give both relative and absolute dates to geological events. They then use these dates to find the rates at which processes occur.

Relative Dating

Methods for relative dating were developed when geology first emerged as a formal science. Geologists still use the following principles today as a means to provide information about geologic history and the timing of geologic events.

The principle of Uniformitarianism states that the geologic processes observed in operation that modify the Earth's crust at present have worked in much the same way over geologic time. A fundamental principle of geology advanced by the 18th century Scottish physician and geologist James Hutton, is that "the present is the key to the past. " In Hutton's words: "the past history of our globe must be explained by what can be seen to be happening now. "

The principle of intrusive relationships concerns crosscutting intrusions. In geology, when an igneous intrusion cuts across a formation of sedimentary rock, it can be determined that the igneous intrusion is younger than the sedimentary rock. There are a number of different types of intrusions, including stocks, laccoliths, batholiths, sills and dikes.

The principle of cross-cutting relationships pertains to the formation of faults and the age of the sequences through which they cut. Faults are younger than the rocks they cut; accordingly, if a fault is found that penetrates some formations but not those on top of it, then the formations that were cut are older than the fault, and the ones that are not cut must be younger than the fault. Finding the key bed in these situations may help determine whether the fault is a normal fault or a thrust fault.

The principle of inclusions and components states that, with sedimentary rocks, if inclusions (or clasts) are found in a formation, then the inclusions must be older than the formation that contains them. For example, in sedimentary rocks, it is common for gravel from an older formation to be ripped up and included in a newer layer. A similar situation with igneous rocks occurs when xenoliths are found. These foreign bodies are picked up as magma or lava flows, and are incorporated, later to cool in the matrix. As a result, xenoliths are older than the rock which contains them. The principle of original horizontality states that the deposition of sediments occurs as essentially horizontal beds. Observation of modern marine and non-marine sediments in a wide variety of environments supports this generalization (although cross-bedding is inclined, the overall orientation of cross-bedded units is horizontal).

The principle of superposition states that a sedimentary rock layer in a tectonically undisturbed sequence is younger than the one beneath it and older than the one above it. Logically a younger layer cannot slip beneath a layer previously deposited. This principle allows sedimentary layers to be viewed as a form of vertical time line, a partial or complete record of the time elapsed from deposition of the lowest layer to deposition of the highest bed.

The principle of faunal succession is based on the appearance of fossils in sedimentary rocks. As organisms exist at the same time period throughout the world, their presence or (sometimes) absence may be used to provide a relative age of the formations in which they are found. Based on principles laid out by William Smith almost a hundred years before the publication of Charles Darwin's theory of evolution, the principles of succession were developed independently of evolutionary thought. The principle becomes quite complex, however, given the uncertainties of fossilization, the localization of fossil types due to lateral changes in habitat (facies change in sedimentary strata), and that not all fossils may be found globally at the same time.

Absolute Dating

Geologists also use methods to determine the absolute age of rock samples and geological events. These dates are useful on their own and may also be used in conjunction with relative dating methods or to calibrate relative methods.

At the beginning of the 20th century, important advancement in geological science was facilitated by the ability to obtain accurate absolute dates to geologic events using radioactive isotopes and other methods. This changed the understanding of geologic time. Previously, geologists could only use fossils and stratigraphic correlation to date sections of rock relative to one another. With isotopic dates it became possible to assign absolute ages to rock units, and these absolute dates could be applied to fossil sequences in which there was datable material, converting the old relative ages into new absolute ages.

For many geologic applications, isotope ratios of radioactive elements are measured in minerals that give the amount of time that has passed since a rock passed through its particular closure temperature, the point at which different radiometric isotopes stop diffusing into and out of the crystal lattice. These are used in geochronologic and thermochronologic studies. Common methods include uranium-lead dating, potassium-argon dating, argon-argon dating and uranium-thorium dating.

These methods are used for a variety of applications. Dating of lava and volcanic ash layers found within a stratigraphic sequence can provide absolute age data for sedimentary rock units which do not contain radioavtive isotopes and calibrate relative dating techniques. These methods can also be used to determine ages of pluton emplacement. Thermochemical techniques can be used to determine temperature profiles within the crust, the uplift of mountain ranges, and paleotopography.

Fractionation of the lanthanide series elements is used to compute ages since rocks were removed from the mantle.

Other methods are used for more recent events. Optically stimulated luminescence and cosmogenic radionucleide dating are used to date surfaces and/or erosion rates. Dendrochronology can also be used for the dating of landscapes. Radiocarbon dating is used for geologically young materials containing organic carbon.

Geological Development of an Area

The geology of an area changes through time as rock units are deposited and inserted and deformational processes change their shapes and locations.

Rock units are first emplaced either by deposition onto the surface or intrusion into the overlying rock. Deposition can occur when sediments settle onto the surface of the Earth and later lithify into sedimentary rock, or when as volcanic material such as volcanic ash or lava flows blanket the surface. Igneous intrusions such as batholiths, laccoliths, dikes, and sills, push upwards into the overlying rock, and crystallize as they intrude.

After the initial sequence of rocks has been deposited, the rock units can be deformed and/or metamorphosed. Deformation typically occurs as a result of horizontal shortening, horizontal extension, or side-to-side (strike-slip) motion. These structural regimes broadly relate to convergent boundaries, divergent boundaries, and transform boundaries, respectively, between tectonic plates.

When rock units are placed under horizontal compression, they shorten and become thicker. Because rock units, other than muds, do not significantly change in volume, this is accomplished in two primary ways: through faulting and folding. In the shallow crust, where brittle deformation can occur, thrust faults form, which cause deeper rock to move on top of shallower rock. Because deeper rock is often older, as noted by the principle of superposition, this can result in older rocks moving on top of younger ones. Movement along faults can result in folding, either because the faults are not planar or because rock layers are dragged along, forming drag folds as slip occurs along the fault. Deeper in the Earth, rocks behave plastically, and fold instead of faulting. These folds can either be those where the material in the center of the fold buckles upwards, creating "antiforms", or where it buckles downwards, creating "synforms". If the tops of the rock units within the folds remain pointing upwards, they are called anticlines and synclines, respectively. If some of the units in the fold are facing downward, the structure is called an overturned anticline or syncline, and if all of the rock units are overturned or the correct up-direction is unknown, they are simply called by the most general terms, antiforms and synforms.

Even higher pressures and temperatures during horizontal shortening can cause both folding and metamorphism of the rocks. This metamorphism causes changes in the mineral composition of the rocks; creates a foliation, or planar surface, that is related to mineral growth under stress. This can remove signs of the original textures of the rocks, such as bedding in sedimentary rocks, flow features of lavas, and crystal patterns in crystalline rocks.

Extension causes the rock units as a whole to become longer and thinner. This is primarily accomplished through normal faulting and through the ductile stretching and thinning. Normal faults drop rock units that are higher below those that are lower. This typically results in younger units being placed below older units. Stretching of units can result in their thinning; in fact, there is a location within the Maria Fold and Thrust Belt in which the entire sedimentary sequence of the Grand Canyon can be seen over a length of less than a meter. Rocks at the depth to be ductilely stretched are often also metamorphosed. These stretched rocks can also pinch into lenses, known as boudins, after the French word for "sausage", because of their visual similarity.

Where rock units slide past one another, strike-slip faults develop in shallow regions, and become shear zones at deeper depths where the rocks deform ductilely.

The addition of new rock units, both depositionally and intrusively, often occurs during deformation. Faulting and other deformational processes result in the creation of topographic gradients, causing material on the rock unit that is increasing in elevation to be eroded by hillslopes and channels. These sediments are deposited on the rock unit that is going down. Continual motion along the fault maintains the topographic gradient in spite of the movement of sediment, and continues to create accommodation space for the material to deposit. Deformational events are often also associated with volcanism and igneous activity. Volcanic ashes and lavas accumulate on the surface, and igneous intrusions enter from below. Dikes, long, planar igneous intrusions, enter along cracks, and therefore often form in large numbers in areas that are being actively deformed. This can result in the emplacement of dike swarms, such as those that are observable across the Canadian shield, or rings of dikes around the lava tube of a volcano.

All of these processes do not necessarily occur in a single environment, and do not necessarily occur in a single order. The Hawaiian Islands, for example, consist almost entirely of layered basaltic lava flows. The sedimentary sequences of the mid-continental United States and the Grand Canyon in the southwestern United States contain almost-undeformed stacks of sedimentary rocks that have remained in place since Cambrian time. Other areas are much more geologically complex. In the southwestern United States, sedimentary, volcanic, and intrusive rocks have been metamorphosed, faulted, foliated, and folded. Even older rocks, such as the Acasta gneiss of the Slave craton in northwestern Canada, the oldest known rock in the world have been metamorphosed to the point where their origin is undiscernable without laboratory analysis. In addition, these processes can occur in stages.

In many places, the Grand Canyon in the southwestern United States being a very visible example, the lower rock units were metamorphosed and deformed, and then deformation ended and the upper, undeformed units were deposited. Although any amount of rock emplacement and rock deformation can occur, and they can occur any number of times, these concepts provide a guide to understanding the geological history of an area.

Methods of Geology

Geologists use a number of field, laboratory, and numerical modeling methods to decipher Earth history and understand the processes that occur on and inside the Earth. In typical geological investigations, geologists use primary information related to petrology (the study of rocks), stratigraphy (the study of sedimentary layers), and structural geology (the study of positions of rock units and their deformation). In many cases, geologists also study modern soils, rivers, landscapes, and glaciers; investigate past and current life and biogeochemical pathways, and use geophysical methods to investigate the subsurface.

Field methods

Geological field work varies depending on the task at hand. Typical fieldwork could consist of:

- ◉ Geological mapping
  - Structural mapping: the locations of major rock units and the faults and folds that led to their placement there.
  - Stratigraphic mapping: the locations of sedimentary facies (lithofacies and biofacies) or the mapping of isopachs of equal thickness of sedimentary rock
  - Surficial mapping: the locations of soils and surficial deposits
- ◉ Surveying of topographic features
  - Creation of topographic maps
  - Work to understand change across landscapes, including:
  - Patterns of erosion and deposition
  - River channel change through migration and avulsion
  - Hillslope processes
- ◉ Subsurface mapping through geophysical methods
  - These methods include:
  - Shallow seismic surveys
  - Ground-penetrating radar
  - Aeromagnetic surveys
  - Electrical resistivity tomography
  - They are used for:
  - Hydrocarbon exploration
  - Finding groundwater
  - Locating buried archaeological artifacts
- ◉ High-resolution stratigraphy
  - Measuring and describing stratigraphic sections on the surface
  - Well drilling and logging
- ◉ Biogeochemistry and geomicrobiology
  - Collecting samples to:
  - Determine biochemical pathways
  - Identify new species of organisms
  - Identify new chemical compounds
  - And to use these discoveries to:

- Understand early life on Earth and how it functioned and metabolized
- Find important compounds for use in pharmaceuticals.
- ◉ Paleontology: excavation of fossil material
  - For research into past life and evolution
  - For museums and education
- ◉ Collection of samples for geochronology and thermochronology
- ◉ Glaciology: measurement of characteristics of glaciers and their motion

Petrology

In addition to identifying rocks in the field, petrologists identify rock samples in the laboratory. Two of the primary methods for identifying rocks in the laboratory are through optical microscopy and by using an electron microprobe. In an optical mineralogy analysis, thin sections of rock samples are analyzed through a petrographic microscope, where the minerals can be identified through their different properties in plane-polarized and cross-polarized light, including their birefringence, pleochroism, twinning, and interference properties with a conoscopic lens. In the electron microprobe, individual locations are analyzed for their exact chemical compositions and variation in composition within individual crystals. Stable and radioactive isotope studies provide insight into the geochemical evolution of rock units.

Petrologists can also use fluid inclusion data and perform high temperature and pressure physical experiments to understand the temperatures and pressures at which different mineral phases appear, and how they change through igneous and metamorphic processes. This research can be extrapolated to the field to understand metamorphic processes and the conditions of crystallization of igneous rocks. This work can also help to explain processes that occur within the Earth, such as subduction and magma chamber evolution.

Structural Geology

Structural geologists use microscopic analysis of oriented thin sections of geologic samples to observe the fabric within the rocks which gives information about strain within the crystalline structure of the rocks. They also plot and combine measurements of geological structures in order to better understand the orientations of faults and folds in order to reconstruct the history of rock deformation in the area. In addition, they perform analog and numerical experiments of rock deformation in large and small settings.

The analysis of structures is often accomplished by plotting the orientations of various features onto stereonets. A stereonet is a stereographic projection of a sphere onto a plane, in which planes are projected as lines and lines are projected as points. These can be used to find the locations of fold axes, relationships between faults, and relationships between other geologic structures.

Among the most well-known experiments in structural geology are those involving orogenic wedges, which are zones in which mountains are built along convergent tectonic plate boundaries. In the analog versions of these experiments, horizontal layers of sand are pulled along a lower surface into a back stop, which results in realistic-looking patterns of faulting and the growth of a critically tapered (all angles remain the same) orogenic wedge. Numerical models work in the same way as these analog models, though they are often more sophisticated and can include patterns of erosion and uplift in the mountain belt. This helps to show the relationship between erosion and the shape of the mountain range. These studies can also give useful information about pathways for metamorphism through pressure, temperature, space, and time.

Stratigraphy

In the laboratory, stratigraphers analyze samples of stratigraphic sections that can be returned from the field, such as those from drill cores. Stratigraphers also analyze data from geophysical surveys that show the locations of stratigraphic units in the subsurface. Geophysical data and well logs can be combined to produce a better view of the subsurface, and stratigraphers often use computer programs to do this in three dimensions. Stratigraphers can then use these data to reconstruct ancient processes occurring on the surface of the Earth, interpret past environments, and locate areas for water, coal, and hydrocarbon extraction.

In the laboratory, biostratigraphers analyze rock samples from outcrop and drill cores for the fossils found in them. These fossils help scientists to date the core and to understand the depositional environment in which the rock units formed. Geochronologists precisely date rocks within the stratigraphic section in order to provide better absolute bounds on the timing and rates of deposition. Magnetic stratigraphers look for signs of magnetic reversals in igneous rock units within the drill cores. Other scientists perform stable isotope studies on the rocks to gain information about past climate.

Planetary geology

With the advent of space exploration in the twentieth century, geologists have begun to look at other planetary bodies in the same ways that have been developed to study the Earth. This new field of study is called planetary geology (sometimes known as astrogeology) and relies on known geologic principles to study other bodies of the solar system.

Although the Greek-language-origin prefix geo refers to Earth, "geology" is often used in conjunction with the names of other planetary bodies when describing their composition and internal processes: examples are "the geology of Mars" and "Lunar geology". Specialised terms such as selenology (studies of the Moon), areology (of Mars), etc., are also in use.

Although planetary geologists are interested in studying all aspects of other planets, a significant focus is to search for evidence of past or present life on other worlds. This has led to many missions whose primary or ancillary purpose is to examine planetary bodies for evidence of life. One of these is the Phoenix lander, which analyzed Martian polar soil for water, chemical, and mineralogical constituents related to biological processes.

APPLIED GEOLOGY

Economic geology

Economic geologists help locate and manage the Earth's natural resources, such as petroleum and coal, as well as mineral resources, which include metals such as iron, copper, and uranium.

Mining geology

Mining geology consists of the extractions of mineral resources from the Earth. Some resources of economic interests include gemstones, metals, and many minerals such as asbestos, perlite, mica, phosphates, zeolites, clay, pumice, quartz, and silica, as well as elements such as sulfur, chlorine, and helium.

Petroleum geology

Petroleum geologists study the locations of the subsurface of the Earth which can contain extractable hydrocarbons, especially petroleum and natural gas. Because many of these reservoirs are found in sedimentary basins, they study the formation of these basins, as well as their sedimentary and tectonic evolution and the present-day positions of the rock units.

Engineering geology

Engineering geology is the application of the geologic principles to engineering practice for the purpose of assuring that the geologic factors affecting the location, design, construction, operation, and maintenance of engineering works are properly addressed. In the field of civil engineering, geological principles and analyses are used in order to ascertain the mechanical principles of the material on which structures are built. This allows tunnels to be built without collapsing, bridges and skyscrapers to be built with sturdy foundations, and buildings to be built that will not settle in clay and mud.

HYDROLOGY AND ENVIRONMENTAL ISSUES

Geology and geologic principles can be applied to various environmental problems such as stream restoration, the restoration of brownfields, and the understanding of the interaction between natural habitat and the geologic environment. Groundwater hydrology, or hydrogeology, is used to locate groundwater, which can often provide a ready supply of uncontaminated water and is especially important in arid regions, and to monitor the spread of contaminants in groundwater wells.

Geologists also obtain data through stratigraphy, boreholes, core samples, and ice cores. Ice cores and sediment cores are used to for paleoclimate reconstructions, which tell geologists about past and present temperature, precipitation, and sea level across the globe. These datasets are our primary source of information on global climate change outside of instrumental data.

History of Geology

The study of the physical material of the Earth dates back at least to ancient Greece when Theophrastus (372–287 BCE) wrote the work Peri Lithon (On Stones). During the Roman period, Pliny the Elder wrote in detail of the many minerals and metals then in practical use – even correctly noting the origin of amber.

Some modern scholars, such as Fielding H. Garrison, are of the opinion that modern geology began in the medieval Islamic world. Abu al-Rayhan al-Biruni (973–1048 CE) was one of the earliest Muslim geologists, whose works included the earliest writings on the geology of India, hypothesizing that the Indian subcontinent was once a sea. Islamic Scholar Ibn Sina (Avicenna, 981–1037) proposed detailed explanations for the formation of mountains, the origin of earthquakes, and other topics central to modern geology, which provided an essential foundation for the later development of the science. In China, the polymath Shen Kuo (1031–1095) formulated a hypothesis for the process of land formation: based on his observation of fossil animal shells in a geological stratum in a mountain hundreds of miles from the ocean, he inferred that the land was formed by erosion of the mountains and by deposition of silt.

Nicolas Steno (1638–1686) is credited with the law of superposition, the principle of original horizontality, and the principle of lateral continuity: three defining principles of stratigraphy.

The word geology was first used by Ulisse Aldrovandi in 1603, then by Jean-André Deluc in 1778 and introduced as a fixed term by Horace-Bénédict de Saussure in 1779. The word is derived from the Greek, meaning "earth" and logos, meaning "speech". But according to another source, the word "geology" comes from a Norwegian, Mikkel Pedersøn Escholt (1600–1699), who was a priest and scholar. Escholt first used the definition in his book titled, Geologica Norvegica (1657). William Smith (1769–1839) drew some of the first geological maps and began the process of ordering rock strata (layers) by examining the fossils contained in them.

James Hutton is often viewed as the first modern geologist. In 1785 he presented a paper entitled Theory of the Earth to the Royal Society of Edinburgh. In his paper, he explained his theory that the Earth must be much older than had previously been supposed in order to allow enough time for mountains to be eroded and for sediments to form new rocks at the bottom of the sea, which in turn were raised up to become dry land. Hutton published a two-volume version of his ideas in 1795 (Vol. 1, Vol. 2).

Followers of Hutton were known as Plutonists because they believed that some rocks were formed by vulcanism, which is the deposition of lava from volcanoes, as opposed to the Neptunists, led by Abraham Werner, who believed that all rocks had settled out of a large ocean whose level gradually dropped over time.

The first geological map of the U. S. was produced in 1809 by William Maclure. In 1807, Maclure commenced the self-imposed task of making a geological survey of the United States. Almost every state in the Union was traversed and mapped by him, the Allegheny Mountains being crossed and recrossed some 50 times. The results of his unaided labours were submitted to the American Philosophical Society in a memoir entitled Observations on the Geology of the United States explanatory of a Geological Map, and published in the Society's Transactions, together with the nation's first geological map. This antedates William Smith's geological map of England by six years, although it was constructed using a different classification of rocks.

Sir Charles Lyell first published his famous book, Principles of Geology, in 1830. This book, which influenced the thought of Charles Darwin, successfully promoted the doctrine of uniformitarianism. This theory states that slow geological processes have occurred throughout the Earth's history and are still occurring today. In contrast, catastrophism is the theory that Earth's features formed in single, catastrophic events and remained unchanged thereafter. Though Hutton believed in uniformitarianism, the idea was not widely accepted at the time.

Much of 19th-century geology revolved around the question of the Earth's exact age. Estimates varied from a few hundred thousand to billions of years. By the early 20th century, radiometric dating allowed the Earth's age to be estimated at two billion years. The awareness of this vast amount of time opened the door to new theories about the processes that shaped the planet.

Some of the most significant advances in 20th-century geology have been the development of the theory of plate tectonics in the 1960s and the refinement of estimates of the planet's age. Plate tectonics theory arose from two separate geological observations: seafloor spreading and continental drift. The theory revolutionized the Earth sciences. Today the Earth is known to be approximately 4. 5 billion years old.

CHAPTER

MINERALOGY | 2

Mineralogy is a subject of geology specializing in the scientific study of chemistry, crystal structure, and physical (including optical) properties of minerals. Specific studies within mineralogy include the processes of mineral origin and formation, classification of minerals, their geographical distribution, as well as their utilization.

HISTORY

Early writing on mineralogy, especially on gemstones, comes from ancient Babylonia, the ancient Greco-Roman world, ancient and medieval China, and Sanskrit texts from ancient India and the ancient Islamic World. Books on the subject included the Naturalis Historia of Pliny the Elder, which not only described many different minerals but also explained many of their properties, and Kitab al Jawahir (Book of Precious Stones) by Persian scientist Al Biruni. The German Renaissance specialist Georgius Agricola wrote works such as De re metallica (On Metals, 1556) and De Natura Fossilium (On the Nature of Rocks, 1546) which began the scientific approach to the subject. Systematic scientific studies of minerals and rocks developed in post-Renaissance Europe. The modern study of mineralogy was founded on the principles of crystallography (the origins of geometric crystallography, itself, can be traced back to the mineralogy practiced in the eighteenth and nineteenth centuries) and to the microscopic study of rock sections with the invention of the microscope in the 17th century.

MODERN MINERALOGY

Historically, mineralogy was heavily concerned with taxonomy of the rock-forming minerals; to this end, the International Mineralogical Association is an organization whose members represent mineralogists in individual countries. Its activities include managing the naming of minerals (via the Commission of New Minerals and Mineral Names), location of known minerals, etc. As of 2004 there are over 4, 000 species of minerals recognized by the IMA. Of these, perhaps 150 can be called "common," another 50 are "occasional," and the rest are "rare" to "extremely rare. "

More recently, driven by advances in experimental technique (such as neutron diffraction) and available computational power, the latter of which has enabled extremely accurate atomic-scale simulations of the behaviour of crystals, the science has branched out to consider more general problems in the fields of inorganic chemistry and solid-state physics. It, however, retains a focus on the crystal structures commonly encountered in rock-forming minerals (such as the perovskites, clay minerals and framework silicates). In particular, the field has made great advances in the understanding of the relationship between the atomic-scale structure of minerals and their function; in nature, prominent examples would be accurate measurement and prediction of the elastic properties of minerals, which has led to new insight into seismological behaviour of rocks and depth-related discontinuities in seismograms of the Earth's mantle. To this end, in their focus on the connection between atomic-scale phenomena and macroscopic properties, the mineral sciences (as they are now commonly known) display perhaps more of an overlap with materials science than any other discipline.

Physical Mineralogy

Physical mineralogy is the specific focus on physical attributes of minerals. Description of physical attributes is the simplest way to identify, classify, and categorize minerals, and they include:

- Crystal structure
- Crystal habit
- Twinning
- Cleavage
- Luster
- Diaphaneity
- Color
- Streak
- Hardness
- Specific gravity

Chemical Mineralogy

Chemical mineralogy focuses on the chemical composition of minerals in order to identify, classify, and categorize them, as well as a means to find beneficial uses from them. There are a few minerals which are classified as whole elements, including sulfur, copper, silver, and gold, yet the vast majority of minerals are chemical compounds, some more complex than others. In terms of major chemical divisions of minerals, most are placed within the isomorphous groups, which are based on analogous chemical composition and similar crystal forms. A good example of isomorphism classification would be the calcite group, containing the minerals calcite, magnesite, siderite, rhodochrosite, and smithsonite.

Biomineralogy

Biomineralogy is a cross-over field between mineralogy, paleontology and biology. It is the study of how plants and animals stabilize minerals under biological control, and the sequencing of mineral replacement of those minerals after deposition. It uses techniques from chemical mineralogy, especially isotopic studies, to determine such things as growth forms in living plants and animals as well as things like the original mineral content of fossils.

Optical Mineralogy

Optical mineralogy is a specific focus of mineralogy that applies sources of light as a means to identify and classify minerals. All minerals which are not part of the cubic system are double refracting, where ordinary light passing through them is broken up into two plane polarized rays that travel at different velocities and refracted at different angles. Mineral substances belonging to the cubic system contain only one index of refraction. Hexagonal and tetragonal mineral substances have two indices, while orthorhombic, monoclinic, and triclinic substances have three indices of refraction. With opaque ore minerals, reflected light from a microscope is needed for identification.

Crystal Structure

X-rays are used to determine the atomic arrangements of minerals and so to identify and classify them. The arrangements of atoms define the crystal structures of the minerals. Some very fine-grained minerals, such as clays, commonly can be identified most readily by their crystal structures. The structure of a mineral also offers a precise way of establishing isomorphism. With knowledge of atomic arrangements and compositions, one may deduce why minerals have specific physical properties, and one may calculate how those properties change with pressure and temperature.

Formation Environments

The environments of mineral formation and growth are highly varied, ranging from slow crystallization at the high temperatures and pressures of igneous melts deep within the Earth's crust to the low temperature precipitation from a saline brine at the Earth's surface.

Various possible methods of formation include:

- Sublimation from volcanic gases
- Deposition from aqueous solutions and hydrothermal brines
- Crystallization from an igneous magma or lava

⊙ Recrystallization due to metamorphic processes and metasomatism

⊙ Crystallization during diagenesis of sediments

⊙ Formation by oxidation and weathering of rocks exposed to the atmosphere or within the soil environment.

Descriptive Mineralogy

Descriptive mineralogy summarizes results of studies performed on mineral substances. It is the scholarly and scientific method of recording the identification, classification, and categorization of minerals, their properties, and their uses. Classifications for descriptive mineralogy includes:

⊙ Native elements

⊙ Sulfides

⊙ Oxides and hydroxides

⊙ Halides

⊙ Carbonates, nitrates and borates

⊙ Sulfates, chromates, molybdates and tungstates

⊙ Phosphates, arsenates and vanadates

⊙ Silicates

⊙ Organic minerals

DETERMINATIVE MINERALOGY

Determinative mineralogy is the actual scientific process of identifying minerals, through data gathering and conclusion. When new minerals are discovered, a standard procedure of scientific analysis is followed, including measures to identify a mineral's formula, its crystallographic data, its optical data, as well as the general physical attributes determined and listed.

Uses

Minerals are essential to various needs within human society, such as minerals used as ores for essential components of metal products used in various commodities and machinery, essential components to building materials such as limestone, marble, granite, gravel, glass, plaster, cement, etc. Minerals are also used in fertilizers to enrich the growth of agricultural crops.

Collecting

Mineral collecting is also a recreational study and collection hobby, with clubs and societies representing the field. Museums, such as the Smithsonian National Museum of Natural History Hall of Geology, Gems, and Minerals, the Natural History Museum of Los Angeles County, the Natural History Museum, London, and the private Mim Mineral Museum in Beirut, Lebanon, have popular collections of mineral specimens on permanent display.

Mineral

A mineral is a naturally occurring substance that is solid and inorganic representable by a chemical formula, usually abiogenic, and has an ordered atomic structure. It is different from a rock, which can be an aggregate of minerals or non-minerals and does not have a specific chemical composition. The exact definition of a mineral is under debate, especially with respect to the requirement a valid species be abiogenic, and to a lesser extent with regard to it having an ordered atomic structure. The study of minerals is called mineralogy.

There are over 4, 900 known mineral species; over 4, 660 of these have been approved by the International Mineralogical Association (IMA). The silicate minerals compose over 90% of the Earth's crust. The diversity and abundance of mineral species is controlled by the Earth's chemistry. Silicon and oxygen constitute approximately 75% of the Earth's crust, which translates directly into the predominance of silicate minerals. Minerals are distinguished by various chemical and physical properties. Differences in chemical composition and crystal structure distinguish various species, and these properties in turn are influenced by the mineral's geological environment of formation. Changes in the temperature, pressure, and bulk composition of a rock mass cause changes in its mineralogy; however, a rock can maintain its bulk composition, but as long as temperature and pressure change, its mineralogy can change as well.

Minerals can be described by various physical properties which relate to their chemical structure and composition. Common distinguishing characteristics include crystal structure and habit, hardness, lustre, diaphaneity, colour, streak, tenacity, cleavage, fracture, parting, and specific gravity. More specific tests for minerals include reaction to acid, magnetism, taste or smell, and radioactivity.

Minerals are classified by key chemical constituents; the two dominant systems are the Dana classification and the Strunz classification. The silicate class of minerals is subdivided into six subclasses by the degree of polymerization in the chemical structure. All silicate minerals have a base unit of a $[SiO4]4$" silica tetrahedra—that is, a silicon cation coordinated by four oxygen anions, which gives the shape of a tetrahedron. These tetrahedra can be polymerized to give the subclasses: orthosilicates (no polymerization, thus single tetrahedra), disilicates (two tetrahedra bonded together), cyclosilicates (rings of tetrahedra), inosilicates (chains of tetrahedra), phyllosilicates (sheets of tetrahedra), and tectosilicates (three-dimensional network of tetrahedra). Other important mineral groups include the native elements, sulfides, oxides, halides, carbonates, sulfates, and phosphates.

DEFINITION

Basic Definition

The general definition of a mineral encompasses the following criteria:

1. Naturally occurring
2. Stable at room temperature
3. Represented by a chemical formula
4. Ordered atomic arrangement
5. Usually abiogenic (not resulting from the activity of living organisms)

The first three general characteristics are less debated than the last two. The first criterion means that a mineral has to form by a natural process, which excludes anthropogenic compounds. Stability at room temperature, in the simplest sense, is synonymous to the mineral being solid. More specifically, a compound has to be stable or metastable at 25 °C. Classical examples of exceptions to this rule include native mercury, which crystallizes at "39 °C, and water ice, which is solid only below 0 °C; as these two minerals were described prior to 1959, they were grandfathered by the International Mineralogical Association (IMA). Modern advances have included extensive study of liquid crystals, which also extensively involve mineralogy. Minerals are chemical compounds, and as such they can be described by fixed or a variable formula. Many mineral groups and species are composed of a solid solution; pure substances are not usually found because of contamination or chemical substitution. For example, the olivine group is described by the variable formula $(Mg, Fe)_2SiO_4$, which is a solid solution of two end-member species, magnesium-rich forsterite and iron-rich fayalite, which are described by a fixed chemical formula. Mineral species themselves could have a variable compositions, such as the sulfide mackinawite, $(Fe, Ni)_9S_8$, which is mostly a ferrous sulfide, but has a very significant nickel impurity that is reflected in its formula.

The requirement of a valid mineral species to be abiogenic has also been described as similar to have to be inorganic; however, this criterion is imprecise and organic compounds have been assigned a separate classification branch. Finally, the requirement of an ordered atomic arrangement is usually synonymous to being crystalline; however, crystals are periodic in addition to being ordered, so the broader criterion is used instead. The presence of an ordered atomic arrangement translates to a variety of macroscopic physical properties, such as crystal form, hardness, and cleavage. There have been several recent proposals to amend the definition to consider biogenic or amorphous substances as minerals. The formal definition of a mineral approved by the IMA in 1995:

"A mineral is an element or chemical compound that is normally crystalline and that has been formed as a result of geological processes. "

In addition, biogenic substances were explicitly excluded:

"Biogenic substances are chemical compounds produced entirely by biological processes without a geological component (e. g., urinary calculi, oxalate crystals in plant tissues, shells of marine molluscs, etc.) and are not regarded as minerals. However, if geological processes were involved in the genesis of the compound, then the product can be accepted as a mineral. "

Recent Advances

Mineral classification schemes and their definitions are evolving to match recent advances in mineral science. Recent changes have included the addition of an organic class, in both the new Dana and the Strunz classification schemes. The organic class includes a very rare group of minerals with hydrocarbons. The IMA Commission on New Minerals and Mineral Names adopted in 2009 a hierarchical scheme for the naming and classification of mineral groups and group names and established seven commissions and four working groups to review and classify minerals into an official listing of their published names. According to these new rules, "mineral species can be grouped in a number of different ways, on the basis of chemistry, crystal structure, occurrence, association, genetic history, or resource, for example, depending on the purpose to be served by the classification. "

The Nickel (1995) exclusion of biogenic substances was not universally adhered to. For example, Lowenstam (1981) stated that "organisms are capable of forming a diverse array of minerals, some of which cannot be formed inorganically in the biosphere. " The distinction is a matter of classification and less to do with the constituents of the minerals themselves. Skinner (2005) views all solids as potential minerals and includes biominerals in the mineral kingdom, which are those that are created by the metabolic activities of organisms. Skinner expanded the previous definition of a mineral to classify "element or compound, amorphous or crystalline, formed through biogeochemical processes, " as a mineral.

Recent advances in high-resolution genetic and x-ray absorption spectroscopy is opening new revelations on the biogeochemical relations between microorganisms and minerals that may make Nickel's (1995) biogenic mineral exclusion obsolete and Skinner's (2005) biogenic mineral inclusion a necessity. For example, the IMA commissioned 'Environmental Mineralogy and Geochemistry Working Group' deals with minerals in the hydrosphere, atmosphere, and biosphere. Mineral forming microorganisms inhabit the areas that this working group deals with. These organisms exist on nearly every rock, soil, and particle surface spanning the globe reaching depths at 1600 metres below the sea floor (possibly further) and 70 kilometres into the stratosphere (possibly entering the mesosphere). Biologists and geologists have started to research and appreciate the magnitude of mineral geoengineering that these creatures are capable of. Bacteria have contributed to the formation of minerals for billions of years and critically define the biogeochemical cycles on this planet. Microorganisms can precipitate metals from solution contributing to the formation of ore deposits in addition to their ability to catalyze mineral dissolution, to respire, precipitate, and form minerals.

Prior to the International Mineralogical Association's listing, over 60 biominerals had been discovered, named, and published. These minerals (a subset tabulated in Lowenstam (1981)) are considered minerals proper according to the Skinner (2005) definition. These biominerals are not listed in the International Mineral Association official list of mineral names, however, many of these biomineral representatives are distributed amongst the 78 mineral classes listed in the Dana classification scheme. Another rare class of minerals (primarily biological in origin) include the mineral liquid crystals that are crystalline and liquid at the same time. To date over 80, 000 liquid crystalline compounds have been identified.

Concerning the use of the term "mineral" to name this family of liquid crystals, one can argue that the term inorganic would be more appropriate. However, inorganic liquid crystals have long been used for organometallic liquid crystals. Therefore, in order to avoid any confusion between these fairly chemically different families, and taking into account that a large number of these liquid crystals occur naturally in nature, we think that the use of the old fashioned but adequate "mineral" adjective taken sensus largo is more specific that an alternative such as "purely inorganic", to name this subclass of the inorganic liquid crystals family.

The Skinner (2005) definition of a mineral takes this matter into account by stating that a mineral can be crystalline or amorphous. Liquid mineral crystals are amorphous. Biominerals and liquid mineral crystals, however, are not the primary form of minerals, most are geological in origin, but these groups do help to identify at the margins of what constitutes a mineral proper. The formal Nickel (1995) definition explicitly mentioned crystalline nature as a key to defining a substance as a mineral. A 2011 article defined icosahedrite, an aluminium-iron-copper alloy as mineral; named for its unique natural icosahedral symmetry, it is also a quasicrystal. Unlike a true crystal, quasicrystals are ordered but not periodic.

Rocks, ores, and gems

Minerals are not equivalent to rocks. Whereas a mineral is a naturally occurring usually solid substance, stable at room temperature, representable by a chemical formula, usually abiogenic, and has an ordered atomic structure, a rock is either an aggregate of one or more minerals, or not composed of minerals at all. Rocks like limestone or quartzite are composed primarily of one mineral—calcite or aragonite in the case of limestone, and quartz in the latter case. Other rocks can be defined by relative abundances of key (essential) minerals; a granite is defined by proportions of quartz, alkali feldspar, and plagioclase feldspar. The other minerals in the rock are termed accessory, and do not greatly affect the bulk composition of the rock. Rocks can also be composed entirely of non-mineral material; coal is a sedimentary rock composed primarily of organically derived carbon.

In rocks, some mineral species and groups are much more abundant than others; these are termed the rock-forming minerals. The major examples of these are quartz, the feldspars, the micas, the amphiboles, the pyroxenes, the olivines, and calcite; except the last one, all of the minerals are silicates. Overall, around 150 minerals are considered particularly important, whether in terms of their abundance or aesthetic value in terms of collecting.

Commercially valuable minerals and rocks are referred to as industrial minerals. For example, muscovite, a white mica, can be used for windows (sometimes referred to as isinglass), as a filler, or as an insulator. Ores are minerals that have a high concentration of a certain element, typically a metal. Examples are cinnabar (HgS), an ore of mercury, sphalerite (ZnS), an ore of zinc, or cassiterite (SnO_2), an ore of tin. Gems are minerals with an ornamental value, and are distinguished from non-gems by their beauty, durability, and usually, rarity. There are about 20 mineral species that qualify as gem minerals, which constitute about 35 of the most common gemstones. Gem minerals are often present in several varieties, and so one mineral can account for several different gemstones; for example, ruby and sapphire are both corundum, Al_2O_3.

NOMENCLATURE AND CLASSIFICATION

In general, a mineral is defined as naturally occurring solid, that is stable at room temperature, representable by a chemical formula, usually abiogenic, and has an ordered atomic structure. However, a mineral can be also narrowed down in terms of a mineral group, series, species, or variety, in order from most broad to least broad. The basic level of definition is that of mineral species, which is distinguished from other species by specific and unique chemical and physical properties. For example, quartz is defined by its formula, $SiO2$, and a specific crystalline structure that distinguishes it from other minerals with the same chemical formula (termed polymorphs). When there exists a range of composition between two minerals species, a mineral series is defined. For example, the biotite series is represented by variable amounts of the endmembers phlogopite, siderophyllite, annite, and eastonite. In contrast, a mineral group is a grouping of mineral species with some common chemical properties that share a crystal structure. The pyroxene group has a common formula of $XY(Si, Al)_2O_6$, where X and Y are both cations, with X typically bigger than Y; the pyroxenes are single-chain silicates that crystallize in either the orthorhombic or monoclinic crystal systems. Finally, a mineral variety is a specific type of mineral species that differs by some physical characteristic, such as colour or crystal habit. An example is amethyst, which is a purple variety of quartz.

Two common classifications are used for minerals; both the Dana and Strunz classifications rely on the composition of the mineral, specifically with regards to important chemical groups, and its structure. The Dana System of Mineralogy was first published in 1837 by James Dwight Dana, a leading geologist of his time; it is in its eighth edition (1997 ed.). The Dana classification, assigns a four-part number to a mineral species. First is its class, based on important compositional groups; next, the type gives the ratio of cations to anions in the mineral; finally, the last two numbers group minerals by structural similarity with a given type or class. The less commonly used Strunz classification, named for German mineralogist Karl Hugo Strunz, is based on the Dana system, but combines both chemical and structural criteria, the latter with regards to distribution of chemical bonds.

There are over 4, 660 approved mineral species. They are most commonly named after a person (45%), followed by discovery location (23%); names based on chemical composition (14%) and physical properties (8%) are the two other major groups of mineral name etymologies. The common suffix -ite of mineral names descends from the ancient Greek suffix - ß ô ç ò (-ites), meaning "connected with or belonging to".

Astrobiology

It has been suggested that biominerals could be important indicators of extraterrestrial life and thus could play an important role in the search for past or present life on the planet Mars. Furthermore, organic components (biosignatures) that are often associated with biominerals are believed to play crucial roles in both pre-biotic and biotic reactions.

On January 24, 2014, NASA reported that current studies by the Curiosity and Opportunity rovers on Mars will now be searching for evidence of ancient life, including a biosphere based on autotrophic, chemotrophic and/or chemolithoautotrophic microorganisms, as well as ancient water, including fluvio-lacustrine environments (plains related to ancient rivers or lakes) that may have been habitable. The search for evidence of habitability, taphonomy (related to fossils), and organic carbon on the planet Mars is now a primary NASA objective.

COMMON ROCK-FORMING MINERALS

3

While rocks consist of aggregates of minerals, minerals themselves are made up of one or a number of chemical elements with a definite chemical composition. Minerals cannot be broken down into smaller units with different chemical compositions in the way that rocks can. More than two thousand three hundred different types of minerals have been identified. Luckily many are rare and the common rocks are made up of a relatively small number of minerals.

IDENTIFYING THE COMMON MINERALS

Since minerals are the building blocks of rocks, it is important that you learn to identify the most common varieties. Minerals can be distinguished using various physical and/or chemical characteristics, but, since chemistry cannot be determined readily in the field, geologists us the physical properties of minerals to identify them. These include features such as crystal form, hardness (relative to a steel blade or you finger nail), colour, lustre, and streak (the colour when a mineral is ground to a powder). More detailed explanations of these terms and other aspects of mineral identification may be found in field handbooks or textbooks. Generally the characteristics listed above can only be determined if the mineral grains are visible in a rock. Thus the identification key distinguishes between rocks in which the grains are visible and those in which the individual mineral components are too small to identify.

The Six Commonest Minerals

The six minerals olivine, quartz, feldspar, mica, pyroxene and amphibole are the commonest rock-forming minerals and are used as important tools in classifying rocks, particularly igneous rocks. Except for quartz, all the minerals listed are actually mineral groups. However, instead of trying to separate all the minerals which make up a group, which is often not possible in the field, they are dealt with here as a single mineral with common characteristics.

Quartz: Quartz is a glassy looking, transparent or translucent mineral which varies in colour from white and grey to smoky. When there are individual crystals they are generally clear, while in larger masses quartz looks more milky white. Quartz is hard - it can easily scratch a steel knife blade. In many rocks, quartz grains are irregular in shape because crystal faces are rare and quartz does not have a cleavage (ie, it does not break on regular flat faces).

Feldspar: Feldspar is the other common, light-coloured rock-forming mineral. Instead of being glassy like quartz, it is generally dull to opaque with a porcelain-like appearance. Colour varies from red, pink, and white (orthoclase) to green, grey and white (plagioclase). Feldspar is also hard but can be scratched by quartz. Feldspar in igneous rocks forms well developed crystals which are roughly rectangular in shape, and they cleave or break along flat faces. The grains, in contrast to quartz, often have straight edges and flat rectangular faces, some of which meet at right angles.

Mica: Mica is easily distinguished by its characteristic of peeling into many thin flat smooth sheets or flakes. This is similar to the cleavage in feldspar except that in the case of mica the cleavage planes are in only one direction and no right angle face joins occur. Mica may be white and pearly (muscovite) or dark and shiny (biotite).

Pyroxene: The most common pyroxene mineral is augite. Augite is generally dark green to black in colour and forms short, stubby crystals which, if you look at an end-on section, have square or rectangular cross-sections.

Amphibole: The most common amphibole is hornblende. Hornblende is quite similar to augite in that both are dark minerals, however hornblende crystals are generally longer, thinner and shinier than augite and the mineral cross-sections are diamond-shaped.

Olivine: Olivine, or peridot in the jewellery trade, is yellow-green, translucent and glassy looking. Crystals are not common; it usually occurs as rounded grains in igneous rocks or as granular masses. Olivine is almost as hard as quartz; it does not have a well-developed cleavage.

Quartz and feldspar are light-coloured minerals; mica, pyroxene, amphibole and olivene are dark-coloured. The colour of a rock will be determined by the proportions of light and dark-coloured minerals present. If most of the grains are quartz and feldspar then the overall appearance of the rock will be light, while the opposite will be true if the minerals are mainly mica, pyroxene, amphibole or olivine. The colour of a rock with between 25 and 50% dark minerals is intermediate.

Other Common rock-forming Minerals

Calcite: Calcite is a very common mineral in sedimentary rocks. It is commonly white to grey in colour. Individual crystals are generally clear and transparent. Calcite is softer than quartz and can be scratched easily by a steel knife blade. In a rock, calcite grains are often irregular to rhomb-like in shape. Calcite's major distinguishing characteristic though is its vigorous reaction with dilute hydrochloric acid. Dolomite is very similar to calcite but does not react well with acid unless powdered first.

Clays: Clay minerals are very fine grained and difficult to tell apart in the field. They can vary in colour from white to grey, brown, red, dark green and black. Clays are plastic and often sticky when wet; they feel smooth when smeared between the fingers.

Magnetite: Magnetite is common in igneous and metamorphic rocks, and some sediments, though usually in only small amounts (1 - 2 %). It is black in colour with a metallic lustre, occurring in small octahedra (like two pyramids stuck together). Easily recognized by its strongly magnetic character.

Pyrite: The commonest of the sulphide minerals, i. e. those minerals containing sulphur as a principle component. It occurs in all rock types, though usually only in small amounts. It is a pale brassy yellow in colour with a metallic lustre and often forms cube-shaped crystals. Also known as "fool's gold".

Talc: Talc occurs in granular or foliated masses sometimes known as soapstone. It is white to green, sometimes grey or brownish. It is very soft and will be scratched by a finger nail. It has a greasy feel.

Rock-forming Minerals

Minerals are the building blocks of rocks. Geologists define a mineral as:

This excludes man-made substances (e. g. synthetic diamonds), organic substances (e. g. chitin), and substances without a fixed composition which are classified as mineraloids.

Some minerals have a definite fixed composition, e. g. quartz is always SiO_2, and calcite is always $CaCO_3$. However, other minerals exhibit a range of compositions between two or more compounds called end-members. For example, plagioclase feldspar has a composition that ranges between end-members anorthite ($CaAl_2Si_2O_8$) and albite ($NaAlSi_3O_8$), so its chemical formula is written as $(Ca, Na)(Al, Si)AlSi_2O_8$.

There are also minerals which form both by inorganic and organic processes. For example, calcite ($CaCO_3$) is a common vein mineral in rocks, and also a shell-forming material in many life forms. Calcite of organic origin conforms to the above definition except for the requirement that it be inorganic. This is an inconsistency in the definition of a mineral that is generally overlooked.

How can a mineral be identified?

A particular mineral can be identified by its unique crystal structure and chemistry. Geologists working in the field, however, don't usually have access to the sophisticated laboratory techniques needed to determine these properties. More commonly, they use Properties which can be observed with the naked eye (or with a hand lens) or determined with simple tools (e. g. a pocket knife).

In hand specimen (a term referring to a piece of rock or mineral able to be easily handled) the nearly constant physical characteristics possessed by a mineral can be used in its identification. It is important to realise that although a particular mineral may be found in several apparently differing forms (e. g. colour, habit), its fundamental physical properties will always be the same.

Useful physical properties for identifying a mineral include its cleavage / fracture, colour, crystal habit / mode of occurrence, hardness, lustre, specific gravity, streak and transparency.

PETROLOGY 4

Petrology is the branch of geology that studies the origin, composition, distribution and structure of rocks.

Lithology was once approximately synonymous with petrography, but in current usage, lithology focuses on macroscopic hand-sample or outcrop-scale description of rocks while petrography is the speciality that deals with microscopic details.

In the petroleum industry, lithology, or more specifically mud logging, is the graphic representation of geological formations being drilled through, and drawn on a log called a mud log. As the cuttings are circulated out of the borehole they are sampled, examined (typically under a 10x microscope) and tested chemically when needed.

METHODOLOGY

Petrology utilizes the classical fields of mineralogy, petrography, optical mineralogy, and chemical analysis to describe the composition and texture of rocks. Modern petrologists also include the principles of geochemistry and geophysics through the study of geochemical trends and cycles and the use of thermodynamic data and experiments in order to better understand the origins of rocks.

Branches

There are three branches of petrology, corresponding to the three types of rocks: igneous, metamorphic, and sedimentary, and another dealing with experimental techniques:

- ◉ Igneous petrology focuses on the composition and texture of igneous rocks (rocks such as granite or basalt which have crystallized from molten rock or magma). Igneous rocks include volcanic and plutonic rocks.

- ◉ Sedimentary petrology focuses on the composition and texture of sedimentary rocks (rocks such as sandstone, shale, or limestone which consist of pieces or particles derived from other rocks or biological or chemical deposits, and are usually bound together in a matrix of finer material).

⦿ Metamorphic petrology focuses on the composition and texture of metamorphic rocks (rocks such as slate, marble, gneiss, or schist which started out as sedimentary or igneous rocks but which have undergone chemical, mineralogical or textural changes due to extremes of pressure, temperature or both)

⦿ Experimental petrology employs high-pressure, high-temperature apparatus to investigate the geochemistry and phase relations of natural or synthetic materials at elevated pressures and temperatures. Experiments are particularly useful for investigating rocks of the lower crust and upper mantle that rarely survive the journey to the surface in pristine condition. They are also one of the prime sources of information about completely inaccessible rocks such as those in the Earth's lower mantle and in the mantles of the other terrestrial planets and the Moon. The work of experimental petrologists has laid a foundation on which modern understanding of igneous and metamorphic processes has been built.

Petrography: is a branch of petrology that focuses on detailed descriptions of rocks. Someone who studies petrography is called a petrographer. The mineral content and the textural relationships within the rock are described in detail. The classification of rocks is based on the information acquired during the petrographic analysis. Petrographic descriptions start with the field notes at the outcrop and include macroscopic description of hand specimens. However, the most important tool for the petrographer is the petrographic microscope. The detailed analysis of minerals by optical mineralogy in thin section and the micro-texture and structure are critical to understanding the origin of the rock. Electron microprobe analysis of individual grains as well as whole rock chemical analysis by atomic absorption or X-ray fluorescence are used in a modern petrographic lab. Individual mineral grains from a rock sample may also be analyzed by X-ray diffraction when optical means are insufficient. Analysis of microscopic fluid inclusions within mineral grains with a heating stage on a petrographic microscope provides clues to the temperature and pressure conditions existent during the mineral formation.

HISTORY

Petrography as a science began in 1828 when Scottish physicist William Nicol invented the technique for producing polarized light by cutting a crystal of Iceland spar, a variety of calcite, into a special prism which became known as the Nicol prism. The addition of two such prisms to the ordinary microscope converted the instrument into a polarizing, or petrographic microscope. Using transmitted light and Nicol prisms, it was possible to determine the internal crystallographic character of very tiny mineral grains, greatly advancing the knowledge of a rock's constituents.

During the 1840s, a development by Henry C. Sorby and others firmly laid the foundation of petrography. This was a technique to study very thin slices of rock. A slice of rock was affixed to a microscope slide and then ground so thin that light could be transmitted through mineral grains that otherwise appeared opaque. The position of adjoining grains was not disturbed, thus permitting analysis of rock texture. Thin section petrography became the standard method of rock study. Since textural details contribute greatly to knowledge of the sequence of crystallization of the various mineral constituents in a rock, petrography progressed into petrogenesis and ultimately into petrology.

It was in Europe, principally in Germany, that petrography advanced in the last half of the nineteenth century.

METHODS OF INVESTIGATION

Macroscopic Characters

The macroscopic characters of rocks, those visible in hand-specimens without the aid of the microscope, are very varied and difficult to describe accurately and fully. The geologist in the field depends principally on them and on a few rough chemical and physical tests; and to the practical engineer, architect and quarry-master they are all-important. Although frequently insufficient in themselves to determine the true nature of a rock, they usually serve for a preliminary classification, and often give all the information needed.

With a small bottle of acid to test for carbonate of lime, a knife to ascertain the hardness of rocks and minerals, and a pocket lens to magnify their structure, the field geologist is rarely at a loss to what group a rock belongs. The fine grained species are often indeterminable in this way, and the minute mineral components of all rocks can usually be ascertained only by microscopic examination. But it is easy to see that a sandstone or grit consists of more or less rounded, water-worn sand grains and if it contains dull, weathered particles of feldspar, shining scales of mica or small crystals of calcite these also rarely escape observation. Shales and clay rocks generally are soft, fine grained, often laminated and not infrequently contain minute organisms or fragments of plants. Limestones are easily marked with a knife-blade, effervesce readily with weak cold acid and often contain entire or broken shells or other fossils. The crystalline nature of a granite or basalt is obvious at a glance, and while the former contains white or pink feldspar, clear vitreous quartz and glancing flakes of mica, the other shows yellow-green olivine, black augite, and gray striated plagioclase.

Other simple tools include the blowpipe (to test the fusibility of detached crystals), the goniometer, the magnet, the magnifying glass and the specific gravity balance.

Microscopic characteristics

When dealing with unfamiliar types or with rocks so fine grained that their component minerals cannot be determined with the aid of a hand lens, a microscope is used. Characteristics observed under the microscope include colour, colour variation under plane polarised light (pleochroism, produced by the lower Nicol prism, or more recently polarising films), fracture characteristics of the grains, refractive index (in comparison to the mounting adhesive, typically Canada Balsam), and optical symmetry (birefringent or isotropic). In toto, these characteristics are sufficient to identify the mineral, and often to quite tightly estimate its major element composition. The process of identifying minerals under the microscope is fairly subtle, but also mechanistic - it would be possible to develop an identification key that would allow a computer to do it. The more difficult and skilful part of optical petrography is identifying the interrelationships between grains and relating them to features seen in hand specimen, at outcrop, or in mapping.

SEPARATION OF COMPONENTS

Separation of the ingredients of a crushed rock powder to obtain pure samples for analysis is a common approach. It may be performed with a powerful, adjustable-strength electromagnet. A weak magnetic field attracts magnetite, then haematite and other iron ores. Silicates that contain iron follow in definite order—biotite, enstatite, augite, hornblende, garnet, and similar ferro-magnesian minerals are successively abstracted. Finally, only the colorless, non-magnetic compounds, such as muscovite, calcite, quartz, and feldspar remain. Chemical methods also are useful.

A weak acid dissolves calcite from crushed limestone, leaving only dolomite, silicates, or quartz. Hydrofluoric acid attacks feldspar before quartz and, if used cautiously, dissolves these and any glassy material in a rock powder before it dissolves augite or hypersthene.

Methods of separation by specific gravity have a still wider application. The simplest of these is levigation—treatment by a current of water. Levigation is extensively employed in mechanical analysis of soils and treatment of ores, but is not so successful with rocks, as their components do not, as a rule, differ greatly in specific gravity. Fluids are used that do not attack most rock-forming minerals, but have a high specific gravity. Solutions of potassium mercuric iodide (sp. gr. 3. 196), cadmium borotungstate (sp. gr. 3. 30), methylene iodide (sp. gr. 3. 32), bromoform (sp. gr. 2. 86), or acetylene bromide (sp. gr. 3. 00) are the principal fluids employed. They may be diluted (with water, benzene, etc.) or concentrated by evaporation.

If the rock is granite consisting of biotite (sp. gr. 3. 1), muscovite (sp. gr. 2. 85), quartz (sp. gr. 2. 65), oligoclase (sp. gr. 2. 64), and orthoclase (sp. gr. 2. 56), the crushed minerals float in methylene iodide. On gradual dilution with benzene they precipitate in the order above. Simple in theory, these methods are tedious in practice, especially as it is common for one rock-making mineral to enclose another. However, expert handling of fresh and suitable rocks yields excellent results.

Chemical Analysis

In addition to naked-eye and microscopic investigation, chemical research methods are of great practical importance to the petrographer. Crushed and separated powders, obtained by the processes above, may be analyzed to determine chemical composition of minerals in the rock qualitatively or quantitatively. Chemical testing, and microscopic examination of minute grains is an elegant and valuable means of discriminating between mineral components of fine-grained rocks.

Thus, the presence of apatite in rock-sections is established by covering a bare rock-section with ammonium molybdate solution. A turbid yellow precipitate forms over the crystals of the mineral in question (indicating the presence of phosphates). Many silicates are insoluble in acids and cannot be tested in this way, but others are partly dissolved, leaving a film of gelatinous silica that can be stained with coloring matters, such as the aniline dyes (nepheline, analcite, zeolites, etc.).

Complete chemical analysis of rocks are also widely used and important, especially in describing new species. Rock analysis has of late years (largely under the influence of the chemical laboratory of the United States Geological Survey) reached a high pitch of refinement and complexity. As many as twenty or twenty-five components may be determined, but for practical purposes a knowledge of the relative proportions of silica, alumina, ferrous and ferric oxides, magnesia, lime, potash, soda and water carry us a long way in determining a rock's position in the conventional classifications.

A chemical analysis is usually sufficient to indicate whether a rock is igneous or sedimentary, and in either case to accurately show what subdivision of these classes it belongs to. In the case of metamorphic rocks it often establishes whether the original mass was a sediment or of volcanic origin.

Specific Gravity

Specific gravity of rocks is determined by use of a balance and pycnometer. It is greatest in rocks containing the most magnesia, iron, and heavy metal while least in rocks rich in alkalis, silica, and water. It diminishes with weathering. Generally, the specific gravity of rocks with the same chemical composition is higher if highly crystalline and lower if wholly or partly vitreous. The specific gravity of the more common rocks range from about 2. 5 to 3. 2.

Archaeological Applications

Archaeologists use petrography to identify mineral components in pottery. This information ties the artifacts to geological areas where the raw materials for the pottery were obtained. In addition to clay, potters often used rock fragments, usually called "temper" or "aplastics", to modify the clay's properties. The geological information obtained from the pottery components provides insight into how potters selected and used local and non-local resources. Archaeologists are able to determine whether pottery found in a particular location was locally produced or traded from elsewhere. This kind of information, along with other evidence, can support conclusions about settlement patterns, group and individual mobility, social contacts, and trade networks. In addition, an understanding of how certain minerals are altered at specific temperatures can allow archaeological petrographers to infer aspects of the ceramic production process itself, such as minimum and maximum temperatures reached during the original firing of the pot.

Ore

An ore is a type of rock that contains sufficient minerals with important elements including metals that can be economically extracted from the rock. The ores are extracted from the earth through mining; they are then refined (often via smelting) to extract the valuable element, or elements.

The grade or concentration of an ore mineral, or metal, as well as its form of occurrence, will directly affect the costs associated with mining the ore. The cost of extraction must thus be weighed against the metal value contained in the rock to determine what ore can be processed and what ore is of too low a grade to be worth mining. Metal ores are generally oxides, sulfides, silicates, or "native" metals (such as native copper) that are not commonly concentrated in the Earth's crust, or "noble" metals (not usually forming compounds) such as gold. The ores must be processed to extract the metals of interest from the waste rock and from the ore minerals. Ore bodies are formed by a variety of geological processes. The process of ore formation is called ore genesis.

ORE DEPOSITS

An ore deposit is an accumulation of ore. This is distinct from a mineral resource as defined by the mineral resource classification criteria. An ore deposit is one occurrence of a particular ore type. Most ore deposits are named according to their location (for example, the Witswatersrand, South Africa), or after a discoverer (e. g. the kambalda nickel shoots are named after drillers), or after some whimsy, a historical figure, a prominent person, something from mythology (phoenix, kraken, serepentleopard, etc.) or the code name of the resource company which found it (e. g. MKD-5 is the in-house name for the Mount Keith nickel).

Classification

Ore deposits are classified according to various criteria developed via the study of economic geology, or ore genesis. The classifications below are typical.

Hydrothermal epigenetic deposits

- Mesothermal lode gold deposits, typified by the Golden Mile, Kalgoorlie
- Archaean conglomerate hosted gold-uranium deposits, typified by Elliot Lake, Canada and Witwatersrand, South Africa
- Carlin–type gold deposits, including;
- Epithermal stockwork vein deposits

Granite related hydrothermal

- IOCG or iron oxide copper gold deposits, typified by the supergiant Olympic Dam Cu-Au-U deposit
- Porphyry copper +/- gold +/- molybdenum +/- silver deposits
- Intrusive-related copper-gold +/- (tin-tungsten), typified by the Tombstone, Arizona deposits
- Hydromagmatic magnetite iron ore deposits and skarns
- Skarn ore deposits of copper, lead, zinc, tungsten, etcetera

Magmatic deposits

- Magmatic nickel-copper-iron-PGE deposits including
  - Cumulate vanadiferous or platinum-bearing magnetite or chromite
  - Cumulate hard-rock titanium (ilmenite) deposits
  - Komatiite hosted Ni-Cu-PGE deposits
  - Subvolcanic feeder subtype, typified by Noril'sk-Talnakh and the Thompson Belt, Canada
  - Intrusive-related Ni-Cu-PGE, typified by Voisey's Bay, Canada and Jinchuan, China
- Lateritic nickel ore deposits, examples include Goro and Acoje, (Philippines) and Ravensthorpe, Western Australia.

Volcanic-related Deposits

- Volcanic hosted massive sulfide (VHMS) Cu-Pb-Zn including;
  - Examples include Teutonic Bore and Golden Grove, Western Australia
  - Besshi type
  - Kuroko type

Metamorphically reworked deposits

⦿ Podiform serpentinite-hosted paramagmatic iron oxide-chromite deposits, typified by Savage River, Tasmania iron ore, Coobina chromite deposit

⦿ Broken Hill Type Pb-Zn-Ag, considered to be a class of reworked SEDEX deposits

Carbonatite-alkaline igneous related

⦿ Phosphorus-tantalite-vermiculite (Phalaborwa South Africa)

⦿ Rare earth elements – Mount Weld, Australia and Bayan Obo, Mongolia

⦿ Diatreme hosted diamond in kimberlite, lamproite or lamprophyre

Sedimentary deposits

⦿ Banded iron formation iron ore deposits, including

    o   Channel-iron deposits or pisolite type iron ore

⦿ Heavy mineral sands ore deposits and other sand dune hosted deposits

⦿ Alluvial gold, diamond, tin, platinum or black sand deposits

⦿ Alluvial oxide zinc deposit type: sole example Skorpion Zinc

Sedimentary hydrothermal deposits

⦿ SEDEX

    • Lead-zinc-silver, typified by Red Dog, McArthur River, Mount Isa, etc.

    • Stratiform arkose-hosted and shale-hosted copper, typified by the Zambian copperbelt.

    • Stratiform tungsten, typified by the Erzgebirge deposits, Czechoslovakia

    • Exhalative spilite-chert hosted gold deposits

⦿ Mississippi valley type (MVT) zinc-lead deposits

⦿ Hematite iron ore deposits of altered banded iron formation

Astrobleme-related ores

⦿ Sudbury Basin nickel and copper, Ontario, Canada

Extraction

The basic extraction of ore deposits follows these steps:

⦿ Prospecting or exploration to find and then define the extent and value of ore where it is located ("ore body")

⦿ Conduct resource estimation to mathematically estimate the size and grade of the deposit

- Conduct a pre-feasibility study to determine the theoretical economics of the ore deposit. This identifies, early on, whether further investment in estimation and engineering studies is warranted and identifies key risks and areas for further work.

- Conduct a feasibility study to evaluate the financial viability, technical and financial risks and robustness of the project and make a decision as whether to develop or walk away from a proposed mine project. This includes mine planning to evaluate the economically recoverable portion of the deposit, the metallurgy and ore recoverability, marketability and payability of the ore concentrates, engineering, milling and infrastructure costs, finance and equity requirements and a cradle to grave analysis of the possible mine, from the initial excavation all the way through to reclamation.

- Development to create access to an ore body and building of mine plant and equipment

- The operation of the mine in an active sense

- Reclamation to make land where a mine had been suitable for future use

CHAPTER

IGNEOUS ROCK AND THEIR
CONSTITUENT MINERALS

5

Igneous rock (derived from the Latin word ignis meaning fire) is one of the three main rock types, the others being sedimentary and metamorphic. Igneous rock is formed through the cooling and solidification of magma or lava. Igneous rock may form with or without crystallization, either below the surface as intrusive (plutonic) rocks or on the surface as extrusive (volcanic) rocks. This magma can be derived from partial melts of pre-existing rocks in either a planet's mantle or crust. Typically, the melting is caused by one or more of three processes: an increase in temperature, a decrease in pressure, or a change in composition. Over 700 types of igneous rocks have been described, most of them having formed beneath the surface of Earth's crust.

GEOLOGICAL SIGNIFICANCE

Igneous and metamorphic rocks make up 90–95% of the top 16 km of the Earth's crust by volume.

Igneous rocks are geologically important because:

⊙ Their minerals and global chemistry give information about the composition of the mantle, from which some igneous rocks are extracted, and the temperature and pressure conditions that allowed this extraction, and/or of other pre-existing rock that melted;

⊙ Their absolute ages can be obtained from various forms of radiometric dating and thus can be compared to adjacent geological strata, allowing a time sequence of events;

⊙ Their features are usually characteristic of a specific tectonic environment, allowing tectonic reconstitutions;

⊙ In some special circumstances they host important mineral deposits (ores): for example, tungsten, tin, and uranium are commonly associated with granites and diorites, whereas ores of chromium and platinum are commonly associated with gabbros.

Morphology and Setting

In terms of modes of occurrence, igneous rocks can be either intrusive (plutonic), extrusive (volcanic) or hypabyssal.

Intrusive

Intrusive igneous rocks are formed from magma that cools and solidifies within the crust of a planet, surrounded by pre-existing rock (called country rock); the magma cools slowly and, as a result, these rocks are coarse grained. The mineral grains in such rocks can generally be identified with the naked eye. Intrusive rocks can also be classified according to the shape and size of the intrusive body and its relation to the other formations into which it intrudes. Typical intrusive formations are batholiths, stocks, laccoliths, sills and dikes. When the magma solidifies within the earth's crust, it cools slowly forming coarse textured rocks, such as granite, gabbro, or diorite.

The central cores of major mountain ranges consist of intrusive igneous rocks, usually granite. When exposed by erosion, these cores (called batholiths) may occupy huge areas of the Earth's surface.

Coarse grained intrusive igneous rocks that form at depth within the crust are termed as abyssal; intrusive igneous rocks that form near the surface are termed hypabyssal.

Extrusive

Extrusive igneous rocks, also known as volcanic rocks, are formed at the crust's surface as a result of the partial melting of rocks within the mantle and crust. Extrusive igneous rocks cool and solidify quicker than intrusive igneous rocks. They are formed by the cooling of molten magma on the earth's surface. The magma, which is brought to the surface through fissures or volcanic eruptions, solidifies at a faster rate. Hence such rocks are smooth, crystalline and fine grained. Basalt is a common extrusive igneous rock and forms lava flows, lava sheets and lava plateaus. Some kinds of basalt solidify to form long polygonal columns. The Giant's Causeway found in Antrim, Northern Ireland is an example.

The melted rock, with or without suspended crystals and gas bubbles, is called magma. It rises because it is less dense than the rock from which it was created. When magma reaches the surface from beneath water or air, it is called lava. Eruptions of volcanoes into air are termed subaerial, whereas those occurring underneath the ocean are termed submarine. Black smokers and mid-ocean ridge basalt are examples of submarine volcanic activity.

The volume of extrusive rock erupted annually by volcanoes varies with plate tectonic setting. Extrusive rock is produced in the following proportions:

- Divergent boundary: 73%
- Convergent boundary (subduction zone): 15%
- Hotspot: 12%.

Magma that erupts from a volcano behaves according to its viscosity, determined by temperature, composition, and crystal content. High-temperature magma, most of which is basaltic in composition, behaves in a manner similar to thick oil and, as it cools, treacle. Long, thin basalt flows with pahoehoe surfaces are common. Intermediate composition magma, such as andesite, tends to form cinder cones of intermingled ash, tuff and lava, and may have a viscosity similar to thick, cold molasses or even rubber when erupted. Felsic magma, such as rhyolite, is usually erupted at low temperature and is up to 10, 000 times as viscous as basalt. Volcanoes with rhyolitic magma commonly erupt explosively, and rhyolitic lava flows are typically of limited extent and have steep margins, because the magma is so viscous.

Felsic and intermediate magmas that erupt often do so violently, with explosions driven by the release of dissolved gases—typically water vapour, but also carbon dioxide. Explosively erupted pyroclastic material is called tephra and includes tuff, agglomerate and ignimbrite. Fine volcanic ash is also erupted and forms ash tuff deposits, which can often cover vast areas.

Because lava cools and crystallizes rapidly, it is fine grained. If the cooling has been so rapid as to prevent the formation of even small crystals after extrusion, the resulting rock may be mostly glass (such as the rock obsidian). If the cooling of the lava happened more slowly, the rocks would be coarse-grained.

Because the minerals are mostly fine-grained, it is much more difficult to distinguish between the different types of extrusive igneous rocks than between different types of intrusive igneous rocks. Generally, the mineral constituents of fine-grained extrusive igneous rocks can only be determined by examination of thin sections of the rock under a microscope, so only an approximate classification can usually be made in the field.

Hypabyssal

Hypabyssal igneous rocks are formed at a depth in between the plutonic and volcanic rocks. These are formed due to cooling and resultant solidification of rising magma just beneath the earth surface. Hypabyssal rocks are less common than plutonic or volcanic rocks and often form dikes, sills, laccoliths, lopoliths, or phacoliths.

Classification

Igneous rocks are classified according to mode of occurrence, texture, mineralogy, chemical composition, and the geometry of the igneous body.

The classification of the many types of different igneous rocks can provide us with important information about the conditions under which they formed. Two important variables used for the classification of igneous rocks are particle size, which largely depends on the cooling history, and the mineral composition of the rock. Feldspars, quartz or feldspathoids, olivines, pyroxenes, amphiboles, and micas are all important minerals in the formation of almost all igneous rocks, and they are basic to the classification of these rocks. All other minerals present are regarded as nonessential in almost all igneous rocks and are called accessory minerals. Types of igneous rocks with other essential minerals are very rare, and these rare rocks include those with essential carbonates.

In a simplified classification, igneous rock types are separated on the basis of the type of feldspar present, the presence or absence of quartz, and in rocks with no feldspar or quartz, the type of iron or magnesium minerals present. Rocks containing quartz (silica in composition) are silica-oversaturated. Rocks with feldspathoids are silica-undersaturated, because feldspathoids cannot coexist in a stable association with quartz.

Igneous rocks that have crystals large enough to be seen by the naked eye are called phaneritic; those with crystals too small to be seen are called aphanitic. Generally speaking, phaneritic implies an intrusive origin; aphanitic an extrusive one.

An igneous rock with larger, clearly discernible crystals embedded in a finer-grained matrix is termed porphyry. Porphyritic texture develops when some of the crystals grow to considerable size before the main mass of the magma crystallizes as finer-grained, uniform material.

Igneous rocks are classified on the basis of texture and composition. Texture refers to the size, shape, and arrangement of the mineral grains or crystals of which the rock is composed.

Texture

Texture is an important criterion for the naming of volcanic rocks. The texture of volcanic rocks, including the size, shape, orientation, and distribution of mineral grains and the intergrain relationships, will determine whether the rock is termed a tuff, a pyroclastic lava or a simple lava.

However, the texture is only a subordinate part of classifying volcanic rocks, as most often there needs to be chemical information gleaned from rocks with extremely fine-grained groundmass or from airfall tuffs, which may be formed from volcanic ash.

Textural criteria are less critical in classifying intrusive rocks where the majority of minerals will be visible to the naked eye or at least using a hand lens, magnifying glass or microscope. Plutonic rocks also tend to be less texturally varied and less prone to gaining structural fabrics. Textural terms can be used to differentiate different intrusive phases of large plutons, for instance porphyritic margins to large intrusive bodies, porphyry stocks and subvolcanic dikes (apophyses). Mineralogical classification is most often used to classify plutonic rocks. Chemical classifications are preferred to classify volcanic rocks, with phenocryst species used as a prefix, e. g. "olivine-bearing picrite" or "orthoclase-phyric rhyolite".

CHEMICAL CLASSIFICATION AND PETROLOGY

Igneous rocks can be classified according to chemical or mineralogical parameters.

Chemical: total alkali-silica content (TAS diagram) for volcanic rock classification used when modal or mineralogic data is unavailable:

- ⊙ Felsic igneous rocks containing a high silica content, greater than 63% SiO_2 (examples granite and rhyolite)

- ⊙ Intermediate igneous rocks containing between 52 – 63% SiO_2 (example andesite and dacite)

- ⊙ Mafic igneous rocks have low silica 45 – 52% and typically high iron – magnesium content (example gabbro and basalt)

- ⊙ Ultramafic rock igneous rocks with less than 45% silica. (examples picrite, komatiite and peridotite)

- ⊙ Alkalic igneous rocks with 5 – 15% alkali (K_2O + Na_2O) content or with a molar ratio of alkali to silica greater than 1:6. (examples phonolite and trachyte)

Chemical classification also extends to differentiating rocks that are chemically similar according to the TAS diagram, for instance;

- ⊙ Ultrapotassic; rocks containing molar K_2O/Na_2O >3

- ⊙ Peralkaline; rocks containing molar (K_2O + Na_2O)/ Al_2O_3 >1

- ⊙ Peraluminous; rocks containing molar (K_2O + Na_2O)/ Al_2O_3<1

An idealized mineralogy (the normative mineralogy) can be calculated from the chemical composition, and the calculation is useful for rocks too fine-grained or too altered for identification of minerals that crystallized from the melt. For instance, normative quartz classifies a rock as silica-oversaturated; an example is rhyolite. In an older terminology, silica oversaturated rocks were called silicic or acidic where the SiO_2 was greater than 66% and the family term quartzolite was applied to the most silicic. A normative feldspathoid classifies a rock as silica-undersaturated; an example is nephelinite.

History of Classification

In 1902, a group of American petrographers proposed that all existing classifications of igneous rocks should be discarded and replaced by a "quantitative" classification based on chemical analysis. They showed how vague, and often unscientific, much of the existing terminology was and argued that as the chemical composition of an igneous rock was its most fundamental characteristic, it should be elevated to prime position. Geological occurrence, structure, mineralogical constitution—the hitherto accepted criteria for the discrimination of rock species—were relegated to the background. The completed rock analysis is first to be interpreted in terms of the rock-forming minerals which might be expected to be formed when the magma crystallizes, e. g., quartz feldspars, olivine, akermannite, Feldspathoids, magnetite, corundum, and so on, and the rocks are divided into groups strictly according to the relative proportion of these minerals to one another.

Mineralogical Classification

For volcanic rocks, mineralogy is important in classifying and naming lavas. The most important criterion is the phenocryst species, followed by the groundmass mineralogy. Often, where the groundmass is aphanitic, chemical classification must be used to properly identify a volcanic rock.

Mineralogic contents – felsic versus mafic

- Felsic rock, highest content of silicon, with predominance of quartz, alkali feldspar and/or feldspathoids: the felsic minerals; these rocks (e. g., granite, rhyolite) are usually light coloured, and have low density.

- Mafic rock, lesser content of silicon relative to felsic rocks, with predominance of mafic minerals pyroxenes, olivines and calcic plagioclase; these rocks (example, basalt, gabbro) are usually dark coloured, and have a higher density than felsic rocks.

- Ultramafic rock, lowest content of silicon, with more than 90% of mafic minerals (e. g., dunite).

For intrusive, plutonic and usually phaneritic igneous rocks (where all minerals are visible at least via microscope), the mineralogy is used to classify the rock. This usually occurs on ternary diagrams, where the relative proportions of three minerals are used to classify the rock.

Example of classification

Granite is an igneous intrusive rock (crystallized at depth), with felsic composition (rich in silica and predominately quartz plus potassium-rich feldspar plus sodium-rich plagioclase) and phaneritic, subeuhedral texture (minerals are visible to the unaided eye and commonly some of them retain original crystallographic shapes).

Magma origination

The Earth's crust averages about 35 kilometers thick under the continents, but averages only some 7–10 kilometers beneath the oceans. The continental crust is composed primarily of sedimentary rocks resting on a crystalline basement formed of a great variety of metamorphic and igneous rocks, including granulite and granite. Oceanic crust is composed primarily of basalt and gabbro. Both continental and oceanic crust rest on peridotite of the mantle.

Rocks may melt in response to a decrease in pressure, to a change in composition (such as an addition of water), to an increase in temperature, or to a combination of these processes.

Other mechanisms, such as melting from a meteorite impact, are less important today, but impacts during the accretion of the Earth led to extensive melting, and the outer several hundred kilometers of our early Earth was probably an ocean of magma. Impacts of large meteorites in the last few hundred million years have been proposed as one mechanism responsible for the extensive basalt magmatism of several large igneous provinces.

DECOMPRESSION

Decompression melting occurs because of a decrease in pressure. The solidus temperatures of most rocks (the temperatures below which they are completely solid) increase with increasing pressure in the absence of water. Peridotite at depth in the Earth's mantle may be hotter than its solidus temperature at some shallower level. If such rock rises during the convection of solid mantle, it will cool slightly as it expands in an adiabatic process, but the cooling is only about 0. 3 °C per kilometer. Experimental studies of appropriate peridotite samples document that the solidus temperatures increase by 3 °C to 4 °C per kilometer. If the rock rises far enough, it will begin to melt. Melt droplets can coalesce into larger volumes and be intruded upwards. This process of melting from the upward movement of solid mantle is critical in the evolution of the Earth.

Decompression melting creates the ocean crust at mid-ocean ridges. It also causes volcanism in intraplate regions, such as Europe, Africa and the Pacific sea floor. There, it is variously attributed either to the rise of mantle plumes (the "Plume hypothesis") or to intraplate extension (the "Plate hypothesis").

Effects of Water and Carbon Dioxide

The change of rock composition most responsible for the creation of magma is the addition of water. Water lowers the solidus temperature of rocks at a given pressure. For example, at a depth of about 100 kilometers, peridotite begins to melt near 800 °C in the presence of excess water, but near or above about 1, 500 °C in the absence of water. Water is driven out of the oceanic lithosphere in subduction zones, and it causes melting in the overlying mantle. Hydrous magmas composed of basalt and andesite are produced directly and indirectly as results of dehydration during the subduction process. Such magmas, and those derived from them, build up island arcs such as those in the Pacific Ring of Fire. These magmas form rocks of the calc-alkaline series, an important part of the continental crust.

The addition of carbon dioxide is relatively a much less important cause of magma formation than the addition of water, but genesis of some silica-undersaturated magmas has been attributed to the dominance of carbon dioxide over water in their mantle source regions. In the presence of carbon dioxide, experiments document that the peridotite solidus temperature decreases by about 200 °C in a narrow pressure interval at pressures corresponding to a depth of about 70 km. At greater depths, carbon dioxide can have more effect: at depths to about 200 km, the temperatures of initial melting of a carbonated peridotite composition were determined to be 450 °C to 600 °C lower than for the same composition with no carbon dioxide. Magmas of rock types such as nephelinite, carbonatite, and kimberlite are among those that may be generated following an influx of carbon dioxide into mantle at depths greater than about 70 km.

Temperature Increase

Increase in temperature is the most typical mechanism for formation of magma within continental crust. Such temperature increases can occur because of the upward intrusion of magma from the mantle. Temperatures can also exceed the solidus of a crustal rock in continental crust thickened by compression at a plate boundary. The plate boundary between the Indian and Asian continental masses provides a well-studied example, as the Tibetan Plateau just north of the boundary has crust about 80 kilometers thick, roughly twice the thickness of normal continental crust. Studies of electrical resistivity deduced from magnetotelluric data have detected a layer that appears to contain silicate melt and that stretches for at least 1, 000 kilometers within the middle crust along the southern margin of the Tibetan Plateau. Granite and rhyolite are types of igneous rock commonly interpreted as products of the melting of continental crust because of increases in temperature. Temperature increases also may contribute to the melting of lithosphere dragged down in a subduction zone.

Magma evolution

Most magmas only entirely melt for small parts of their histories. More typically, they are mixes of melt and crystals, and sometimes also of gas bubbles. Melt, crystals, and bubbles usually have different densities, and so they can separate as magmas evolve.

As magma cools, minerals typically crystallize from the melt at different temperatures (fractional crystallization). As minerals crystallize, the composition of the residual melt typically changes. If crystals separate from the melt, then the residual melt will differ in composition from the parent magma. For instance, a magma of gabbroic composition can produce a residual melt of granitic composition if early formed crystals are separated from the magma. Gabbro may have a liquidus temperature near 1, 200 °C, and the derivative granite-composition melt may have a liquidus temperature as low as about 700 °C. Incompatible elements are concentrated in the last residues of magma during fractional crystallization and in the first melts produced during partial melting: either process can form the magma that crystallizes to pegmatite, a rock type commonly enriched in incompatible elements. Bowen's reaction series is important for understanding the idealised sequence of fractional crystallisation of a magma.

Magma composition can be determined by processes other than partial melting and fractional crystallization. For instance, magmas commonly interact with rocks they intrude, both by melting those rocks and by reacting with them. Magmas of different compositions can mix with one another. In rare cases, melts can separate into two immiscible melts of contrasting compositions.

There are relatively few minerals that are important in the formation of common igneous rocks, because the magma from which the minerals crystallize is rich in only certain elements: silicon, oxygen, aluminium, sodium, potassium, calcium, iron, and magnesium. These are the elements that combine to form the silicate minerals, which account for over ninety percent of all igneous rocks. The chemistry of igneous rocks is expressed differently for major and minor elements and for trace elements. Contents of major and minor elements are conventionally expressed as weight percent oxides (e. g., 51% SiO_2, and 1. 50% TiO_2). Abundances of trace elements are conventionally expressed as parts per million by weight (e. g., 420 ppm Ni, and 5. 1 ppm Sm). The term "trace element" is typically used for elements present in most rocks at abundances less than 100 ppm or so, but some trace elements may be present in some rocks at abundances exceeding 1, 000 ppm. The diversity of rock compositions has been defined by a huge mass of analytical data—over 230, 000 rock analyses can be accessed on the web through a site sponsored by the U. S. National Science Foundation.

CHAPTER 6

SEDIMENTARY ROCKS

Sedimentary rocks are types of rock that are formed by the deposition of material at the Earth's surface and within bodies of water. Sedimentation is the collective name for processes that cause mineral and/or organic particles (detritus) to settle and accumulate or minerals to precipitate from a solution. Particles that form a sedimentary rock by accumulating are called sediment. Before being deposited, sediment was formed by weathering and erosion in a source area, and then transported to the place of deposition by water, wind, ice, mass movement or glaciers which are called agents of denudation.

The sedimentary rock cover of the continents of the Earth's crust is extensive, but the total contribution of sedimentary rocks is estimated to be only 8% of the total volume of the crust. Sedimentary rocks are only a thin veneer over a crust consisting mainly of igneous and metamorphic rocks. Sedimentary rocks are deposited in layers as strata, forming a structure called bedding. The study of sedimentary rocks and rock strata provides information about the subsurface that is useful for civil engineering, for example in the construction of roads, houses, tunnels, canals or other structures. Sedimentary rocks are also important sources of natural resources like coal, fossil fuels, drinking water or ores. The study of the sequence of sedimentary rock strata is the main source for scientific knowledge about the Earth's history, including palaeogeography, paleoclimatology and the history of life. The scientific discipline that studies the properties and origin of sedimentary rocks is called sedimentology. Sedimentology is part of both geology and physical geography and overlaps partly with other disciplines in the Earth sciences, such as pedology, geomorphology, geochemistry and structural geology.

GENETIC CLASSIFICATION

Based on the processes responsible for their formation, sedimentary rocks can be subdivided into four groups: clastic sedimentary rocks, biochemical (or biogenic) sedimentary rocks, chemical sedimentary rocks and a fourth category for "other" sedimentary rocks formed by impacts, volcanism, and other minor processes.

Clastic Sedimentary Rocks

Clastic sedimentary rocks are composed of silicate minerals and rock fragments that were transported by moving fluids (as bed load, suspended load, or by sediment gravity flows) and were deposited when these fluids came to rest. Clastic rocks are composed largely of quartz, feldspar, rock (lithic) fragments, clay minerals, and mica; numerous other minerals may be present as accessories and may be important locally.

Clastic sediment, and thus clastic sedimentary rocks, are subdivided according to the dominant particle size (diameter). Most geologists use the Udden-Wentworth grain size scale and divide unconsolidated sediment into three fractions: gravel (>2 mm diameter), sand (1/16 to 2 mm diameter), and mud (clay is <1/256 mm and silt is between 1/16 and 1/256 mm). The classification of clastic sedimentary rocks parallels this scheme; conglomerates and breccias are made mostly of gravel, sandstones are made mostly of sand, and mudrocks are made mostly of mud. This tripartite subdivision is mirrored by the broad categories of rudites, arenites, and lutites, respectively, in older literature.

Subdivision of these three broad categories is based on differences in clast shape (conglomerates and breccias), composition (sandstones), grain size and/or texture (mudrocks).

Conglomerates and Breccias

Conglomerates are dominantly composed of rounded gravel and breccias are composed of dominantly angular gravel.

Sandstones

Sandstone classification schemes vary widely, but most geologists have adopted the Dott scheme, which uses the relative abundance of quartz, feldspar, and lithic framework grains and the abundance of muddy matrix between these larger grains.

Composition of Framework Grains

The relative abundance of sand-sized framework grains determines the first word in a sandstone name. For naming purposes, the abundance of framework grains is normalized to quartz, feldspar, and lithic fragments formed from other rocks. These are the three most abundant components of sandstones; all other minerals are considered accessories and not used in the naming of the rock, regardless of abundance.

- ⊙ Quartz sandstones have >90% quartz grains

- ⊙ Feldspathic sandstones have <90% quartz grains and more feldspar grains than lithic grains

- ⊙ Lithic sandstones have <90% quartz grains and more lithic grains than feldspar grains

Abundance of Muddy Matrix between Sand Grains

When sand-sized particles are deposited, the space between the sand grains either remains open or is filled with mud (silt and/or clay sized particle).

- ⦿ "Clean" sandstones with open pore space (that may later be filled with cement) are called arenites

- ⦿ Muddy sandstones with abundant (>10%) muddy matrix are called wackes.

Six sandstone names are possible using descriptors for grain composition (quartz-, feldspathic-, and lithic-) and amount of matrix (wacke or arenite). For example, a quartz arenite would be composed of mostly (>90%) quartz grains and have little/no clayey matrix between the grains, a lithic wacke would have abundant lithic grains (<90% quartz, remainder would have more lithics than feldspar) and abundant muddy matrix, etc.

Although the Dott classification scheme is widely used by sedimentologists, common names like greywacke, arkose, and quartz sandstone are still widely used by nonspecialists and in popular literature.

Mudrocks

Mudrocks are sedimentary rocks composed of at least 50% silt- and clay-sized particles. These relatively fine-grained particles are commonly transported as suspended particles by turbulent flow in water or air, and deposited as the flow calms and the particles settle out of suspension.

Most authors presently use the term "mudrock" to refer to all rocks composed dominantly of mud. Mudrocks can be divided into siltstones (composed dominantly of silt-sized particles), mudstones (subequal mixture of silt- and clay-sized particles), and claystones (composed mostly of clay-sized particles). Most authors use "shale" as a term for a fissile mudrock (regardless of grain size) although some older literature uses the term "shale" as a synonym for mudrock.

BIOCHEMICAL SEDIMENTARY ROCKS

Biochemical sedimentary rocks are created when organisms use materials dissolved in air or water to build their tissue. Examples include:

- ⦿ Most types of limestone are formed from the calcareous skeletons of organisms such as corals, mollusks, and foraminifera.

- ⦿ Coal which forms as plants remove carbon from the atmosphere and combine with other elements to build their tissue.

- ⦿ Deposits of chert formed from the accumulation of siliceous skeletons from microscopic organisms such as radiolaria and diatoms.

Chemical sedimentary rocks

Chemical sedimentary rock forms when mineral constituents in solution become supersaturated and inorganically precipitate. Common chemical sedimentary rocks include oolitic limestone and rocks composed of evaporite minerals such as halite (rock salt), sylvite, barite and gypsum.

"Other" sedimentary rocks

This fourth miscellaneous category includes rocks formed by Pyroclastic flows, impact breccias, volcanic breccias, and other relatively uncommon processes.

Compositional classification schemes

Alternatively, sedimentary rocks can be subdivided into compositional groups based on their mineralogy:

- ◉ Siliciclastic sedimentary rocks, as described above, are dominantly composed of silicate minerals. The sediment that makes up these rocks was transported as bed load, suspended load, or by sediment gravity flows. Siliciclastic sedimentary rocks are subdivided into conglomerates and breccias, sandstone, and mudrocks.

- ◉ Carbonate sedimentary rocks are composed of calcite (rhombohedral $CaCO_3$), aragonite (orthorhombic $CaCO_3$), dolomite ($CaMg(CO_3)_2$), and other carbonate minerals based on the $CO_2"3$ ion. Common examples include limestone and dolostone.

- ◉ Evaporite sedimentary rocks are composed of minerals formed from the evaporation of water. The most common evaporite minerals are carbonates (calcite and others based on $CO_2"3$), chlorides (halite and others built on $Cl"$), and sulfates (gypsum and others built on $SO_2"4$). Evaporite rocks commonly include abundant halite (rock salt), gypsum, and anhydrite.

- ◉ Organic-rich sedimentary rocks have significant amounts of organic material, generally in excess of 3% total organic carbon. Common examples include coal, oil shale as well as source rocks for oil and natural gas.

- ◉ Siliceous sedimentary rocks are almost entirely composed of silica (SiO_2), typically as chert, opal, chalcedony or other microcrystalline forms.

- ◉ Iron-rich sedimentary rocks are composed of >15% iron; the most common forms are banded iron formations and ironstones

- ◉ Phosphatic sedimentary rocks are composed of phosphate minerals and contain more than 6. 5% phosphorus; examples include deposits of phosphate nodules, bone beds, and phosphatic mudrocks

DEPOSITION AND DIAGENESIS

Sediment Transport and Deposition

Sedimentary rocks are formed when sediment is deposited out of air, ice, wind, gravity, or water flows carrying the particles in suspension. This sediment is often formed when weathering and erosion break down a rock into loose material in a source area. The material is then transported from the source area to the deposition area. The type of sediment transported depends on the geology of the hinterland (the source area of the sediment). However, some sedimentary rocks, like evaporites, are composed of material that formed at the place of deposition. The nature of a sedimentary rock, therefore, not only depends on sediment supply, but also on the sedimentary depositional environment in which it formed.

Diagenesis

The term diagenesis is used to describe all the chemical, physical, and biological changes, including cementation, undergone by a sediment after its initial deposition, exclusive of surface weathering. Some of these processes cause the sediment to consolidate: a compact, solid substance forms out of loose material. Young sedimentary rocks, especially those of Quaternary age (the most recent period of the geologic time scale) are often still unconsolidated. As sediment deposition builds up, the overburden (or lithostatic) pressure rises, and a process known as lithification takes place.

Sedimentary rocks are often saturated with seawater or groundwater, in which minerals can dissolve or from which minerals can precipitate. Precipitating minerals reduce the pore space in a rock, a process called cementation. Due to the decrease in pore space, the original connate fluids are expelled. The precipitated minerals form a cement and make the rock more compact and competent. In this way, loose clasts in a sedimentary rock can become "glued" together.

When sedimentation continues, an older rock layer becomes buried deeper as a result. The lithostatic pressure in the rock increases due to the weight of the overlying sediment. This causes compaction, a process in which grains mechanically reorganize. Compaction is, for example, an important diagenetic process in clay, which can initially consist of 60% water. During compaction, this interstitial water is pressed out of pore spaces. Compaction can also be the result of dissolution of grains by pressure solution. The dissolved material precipitates again in open pore spaces, which means there is a net flow of material into the pores. However, in some cases a certain mineral dissolves and does not precipitate again. This process, called leaching, increases pore space in the rock.

Some biochemical processes, like the activity of bacteria, can affect minerals in a rock and are therefore seen as part of diagenesis. Fungi and plants (by their roots) and various other organisms that live beneath the surface can also influence diagenesis.

Burial of rocks due to ongoing sedimentation leads to increased pressure and temperature, which stimulates certain chemical reactions. An example is the reactions by which organic material becomes lignite or coal. When temperature and pressure increase still further, the realm of diagenesis makes way for metamorphism, the process that forms metamorphic rock.

PROPERTIES

Color

The color of a sedimentary rock is often mostly determined by iron, an element with two major oxides: iron(II) oxide and iron(III) oxide. Iron(II) oxide only forms under anoxic circumstances and gives the rock a grey or greenish colour. Iron(III) oxide is often in the form of the mineral hematite and gives the rock a reddish to brownish colour. In arid continental climates rocks are in direct contact with the atmosphere, and oxidation is an important process, giving the rock a red or orange colour. Thick sequences of red sedimentary rocks formed in arid climates are called red beds. However, a red colour does not necessarily mean the rock formed in a continental environment or arid climate.

The presence of organic material can colour a rock black or grey. Organic material is in nature formed from dead organisms, mostly plants. Normally, such material eventually decays by oxidation or bacterial activity. Under anoxic circumstances, however, organic material cannot decay and becomes a dark sediment, rich in organic material. This can, for example, occur at the bottom of deep seas and lakes. There is little water current in such environments, so oxygen from surface water is not brought down, and the deposited sediment is normally a fine dark clay. Dark rocks rich in organic material are therefore often shales.

Texture

The size, form and orientation of clasts or minerals in a rock is called its texture. The texture is a small-scale property of a rock, but determined many of its large-scale properties, such as the density, porosity or permeability.

Clastic rocks have a 'clastic texture', which means they consist of clasts. The 3D orientation of these clasts is called the fabric of the rock. Between the clasts, the rock can be composed of a matrix or a cement (the latter can consist of crystals of one or more precipitated minerals). The size and form of clasts can be used to determine the velocity and direction of current in the sedimentary environment where the rock was formed; fine, calcareous mud only settles in quiet water while gravel and larger clasts are only deposited by rapidly moving water. The grain size of a rock is usually expressed with the Wentworth scale, though alternative scales are sometimes used.

The grain size can be expressed as a diameter or a volume, and is always an average value – a rock is composed of clasts with different sizes. The statistical distribution of grain sizes is different for different rock types and is described in a property called the sorting of the rock. When all clasts are more or less of the same size, the rock is called 'well-sorted', and when there is a large spread in grain size, the rock is called 'poorly sorted'.

The form of clasts can reflect the origin of the rock.

Coquina, a rock composed of clasts of broken shells, can only form in energetic water. The form of a clast can be described by using four parameters:

- Surface texture describes the amount of small-scale relief of the surface of a grain that is too small to influence the general shape.

- Rounding describes the general smoothness of the shape of a grain.

- 'Sphericity' describes the degree to which the grain approaches a sphere.

- 'Grain form' describes the three dimensional shape of the grain.

Chemical sedimentary rocks have a non-clastic texture, consisting entirely of crystals. To describe such a texture, only the average size of the crystals and the fabric are necessary.

Mineralogy

Most sedimentary rocks contain either quartz (especially siliciclastic rocks) or calcite (especially carbonate rocks). In contrast to igneous and metamorphic rocks, a sedimentary rock usually contains very few different major minerals. However, the origin of the minerals in a sedimentary rock is often more complex than in an igneous rock. Minerals in a sedimentary rock can have formed by precipitation during sedimentation or by diagenesis. In the second case, the mineral precipitate can have grown over an older generation of cement. A complex diagenetic history can be studied by optical mineralogy, using a petrographic microscope.

Carbonate rocks dominantly consist of carbonate minerals like calcite, aragonite or dolomite. Both cement and clasts (including fossils and ooids) of a carbonate rock can consist of carbonate minerals. The mineralogy of a clastic rock is determined by the supplied material from the source area, the manner of transport to the place of deposition and the stability of a particular mineral. The stability of the major rock forming minerals (their resistance to weathering) is expressed by Bowen's reaction series. In this series, quartz is the most stable, followed by feldspar, micas, and other less stable minerals that are only present when little weathering has occurred. The amount of weathering depends mainly on the distance to the source area, the local climate and the time it took for the sediment to be transported there. In most sedimentary rocks, mica, feldspar and less stable minerals have reacted to clay minerals like kaolinite, illite or smectite.

Fossils

Among the three major types of rock, fossils are most commonly found in sedimentary rock. Unlike most igneous and metamorphic rocks, sedimentary rocks form at temperatures and pressures that do not destroy fossil remnants. Often these fossils may only be visible when studied under a microscope (microfossils) or with a loupe.

Dead organisms in nature are usually quickly removed by scavengers, bacteria, rotting and erosion, but sedimentation can contribute to exceptional circumstances where these natural processes are unable to work, causing fossilisation. The chance of fossilisation is higher when the sedimentation rate is high (so that a carcass is quickly buried), in anoxic environments (where little bacterial activity occurs) or when the organism had a particularly hard skeleton. Larger, well-preserved fossils are relatively rare.

Fossils can be both the direct remains or imprints of organisms and their skeletons. Most commonly preserved are the harder parts of organisms such as bones, shells, and the woody tissue of plants. Soft tissue has a much smaller chance of being preserved and fossilized, and soft tissue of animals older than 40 million years is very rare. Imprints of organisms made while still alive are called trace fossils. Examples are burrows, footprints, etc.

Being part of a sedimentary or metamorphic rock, fossils undergo the same diagenetic processes as rock. A shell consisting of calcite can, for example, dissolve while a cement of silica then fills the cavity. In the same way, precipitating minerals can fill cavities formerly occupied by blood vessels, vascular tissue or other soft tissues. This preserves the form of the organism but changes the chemical composition, a process called permineralization. The most common minerals in permineralization cements are carbonates (especially calcite), forms of amorphous silica (chalcedony, flint, chert) and pyrite. In the case of silica cements, the process is called lithification.

At high pressure and temperature, the organic material of a dead organism undergoes chemical reactions in which volatiles like water and carbon dioxide are expulsed. The fossil, in the end, consists of a thin layer of pure carbon or its mineralized form, graphite. This form of fossilisation is called carbonisation. It is particularly important for plant fossils. The same process is responsible for the formation of fossil fuels like lignite or coal.

PRIMARY SEDIMENTARY STRUCTURES

Structures in sedimentary rocks can be divided into 'primary' structures (formed during deposition) and 'secondary' structures (formed after deposition). Unlike textures, structures are always large-scale features that can easily be studied in the field. Sedimentary structures can indicate something about the sedimentary environment or can serve to tell which side originally faced up where tectonics have tilted or overturned sedimentary layers.

Sedimentary rocks are laid down in layers called beds or strata. A bed is defined as a layer of rock that has a uniform lithology and texture. Beds form by the deposition of layers of sediment on top of each other. The sequence of beds that characterizes sedimentary rocks is called bedding. Single beds can be a couple of centimetres to several meters thick. Finer, less pronounced layers are called laminae, and the structure it forms in a rock is called lamination. Laminae are usually less than a few centimetres thick. Though bedding and lamination are often originally horizontal in nature, this is not always the case. In some environments, beds are deposited at a (usually small) angle. Sometimes multiple sets of layers with different orientations exist in the same rock, a structure called cross-bedding. Cross-bedding forms when small-scale erosion occurs during deposition, cutting off part of the beds. Newer beds then form at an angle to older ones.

The opposite of cross-bedding is parallel lamination, where all sedimentary layering is parallel. With laminations, differences are generally caused by cyclic changes in the sediment supply, caused, for example, by seasonal changes in rainfall, temperature or biochemical activity. Laminae that represent seasonal changes (similar to tree rings) are called varves. Any sedimentary rock composed of millimeter or finer scale layers can be named with the general term laminite. Some rocks have no lamination at all; their structural character is called massive bedding.

Graded bedding is a structure where beds with a smaller grain size occur on top of beds with larger grains. This structure forms when fast flowing water stops flowing. Larger, heavier clasts in suspension settle first, then smaller clasts. Though graded bedding can form in many different environments, it is characteristic for turbidity currents.

The bedform (the surface of a particular bed) can be indicative for a particular sedimentary environment, too. Examples of bed forms include dunes and ripple marks. Sole markings, such as tool marks and flute casts, are groves dug into a sedimentary layer that are preserved. These are often elongated structures and can be used to establish the direction of the flow during deposition.

Ripple marks also form in flowing water. There are two types: asymmetric wave ripples and symmetric current ripples. Environments where the current is in one direction, such as rivers, produce asymmetric ripples. The longer flank of such ripples is oriented opposite to the direction of the current. Wave ripples occur in environments where currents occur in all directions, such as tidal flats.

Mudcracks are a bed form caused by the dehydration of sediment that occasionally comes above the water surface. Such structures are commonly found at tidal flats or point bars along rivers.

Secondary Sedimentary Structures

Secondary sedimentary structures are structures in sedimentary rocks which formed after deposition. Such structures form by chemical, physical and biological processes inside the sediment. They can be indicators for circumstances after deposition. Some can be used as way up criteria.

Organic presence in a sediment can leave more traces than just fossils. Preserved tracks and burrows are examples of trace fossils (also called ichnofossils). Some trace fossils such as paw prints of dinosaurs or early humans can capture human imagination, but such traces are relatively rare. Most trace fossils are burrows of molluscs or arthropods. This burrowing is called bioturbation by sedimentologists. It can be a valuable indicator of the biological and ecological environment after the sediment was deposited. On the other hand, the burrowing activity of organisms can destroy other (primary) structures in the sediment, making a reconstruction more difficult.

Secondary structures can also have been formed by diagenesis or the formation of a soil (pedogenesis) when a sediment is exposed above the water level. An example of a diagenetic structure common in carbonate rocks is a stylolite. Stylolites are irregular planes where material was dissolved into the pore fluids in the rock. The result of precipitation of a certain chemical species can be colouring and staining of the rock, or the formation of concretions. Concretions are roughly concentric bodies with a different composition from the host rock. Their formation can be the result of localized precipitation due to small differences in composition or porosity of the host rock, such as around fossils, inside burrows or around plant roots. In carbonate rocks such as limestone or chalk, chert or flint concretions are common, while terrestrial sandstones can have iron concretions. Calcite concretions in clay are called septarian concretions.

After deposition, physical processes can deform the sediment, forming a third class of secondary structures. Density contrasts between different sedimentary layers, such as between sand and clay, can result in flame structures or load casts, formed by inverted diapirism. The diapirism causes the denser upper layer to sink into the other layer. Sometimes, density contrast can result or grow when one of the lithologies dehydrates. Clay can be easily compressed as a result of dehydration, while sand retains the same volume and becomes relatively less dense. On the other hand, when the pore fluid pressure in a sand layer surpasses a critical point, the sand can flow through overlying clay layers, forming discordant bodies of sedimentary rock called sedimentary dykes (the same process can form mud volcanoes on the surface).

A sedimentary dyke can also be formed in a cold climate where the soil is permanently frozen during a large part of the year. Frost weathering can form cracks in the soil that fill with rubble from above. Such structures can be used as climate indicators as well as way up structures.

Density contrasts can also cause small-scale faulting, even while sedimentation goes on (syn-sedimentary faulting). Such faulting can also occur when large masses of non-lithified sediment are deposited on a slope, such as at the front side of a delta or the continental slope. Instabilities in such sediments can result in slumping. The resulting structures in the rock are syn-sedimentary folds and faults, which can be difficult to distinguish from folds and faults formed by tectonic forces in lithified rocks.

Sedimentary Environments

The setting in which a sedimentary rock forms is called the sedimentary environment. Every environment has a characteristic combination of geologic processes and circumstances. The type of sediment that is deposited is not only dependent on the sediment that is transported to a place, but also on the environment itself.

A marine environment means the rock was formed in a sea or ocean. Often, a distinction is made between deep and shallow marine environments. Deep marine usually refers to environments more than 200 m below the water surface. Shallow marine environments exist adjacent to coastlines and can extend out to the boundaries of the continental shelf. The water in such environments has a generally higher energy than that in deep environments, because of wave activity. This means coarser sediment particles can be transported and the deposited sediment can be coarser than in deep environments. When the available sediment is transported from the continent, an alternation of sand, clay and silt is deposited. When the continent is far away, the amount of such sediment brought in may be small, and biochemical processes dominate the type of rock that forms. Especially in warm climates, shallow marine environments far offshore mainly see deposition of carbonate rocks. The shallow, warm water is an ideal habitat for many small organisms that build carbonate skeletons. When these organisms die their skeletons sink to the bottom, forming a thick layer of calcareous mud that may lithify into limestone. Warm shallow marine environments also are ideal environments for coral reefs, where the sediment consists mainly of the calcareous skeletons of larger organisms.

In deep marine environments, the water current over the sea bottom is small. Only fine particles can be transported to such places. Typically sediments depositing on the ocean floor are fine clay or small skeletons of micro-organisms. At 4 km depth, the solubility of carbonates increases dramatically (the depth zone where this happens is called the lysocline). Calcareous sediment that sinks below the lysocline dissolve, so no limestone can be formed below this depth. Skeletons of micro-organisms formed of silica (such as radiolarians) still deposit though. An example of a rock formed out of silica skeletons is radiolarite. When the bottom of the sea has a small inclination, for example at the continental slopes, the sedimentary cover can become unstable, causing turbidity currents. Turbidity currents are sudden disturbances of the normally quite deep marine environment and can cause the geologically speaking instantaneous deposition of large amounts of sediment, such as sand and silt. The rock sequence formed by a turbidity current is called a turbidite.

The coast is an environment dominated by wave action. At the beach, dominantly coarse sediment like sand or gravel is deposited, often mingled with shell fragments. Tidal flats and shoals are places that sometimes dry out because of the tide. They are often cross-cut by gullies, where the current is strong and the grain size of the deposited sediment is larger. Where along a coast (either the coast of a sea or a lake) rivers enter the body of water, deltas can form. These are large accumulations of sediment transported from the continent to places in front of the mouth of the river. Deltas are dominantly composed of clastic sediment.

A sedimentary rock formed on the land has a continental sedimentary environment. Examples of continental environments are lagoons, lakes, swamps, floodplains and alluvial fans. In the quiet water of swamps, lakes and lagoons, fine sediment is deposited, mingled with organic material from dead plants and animals. In rivers, the energy of the water is much higher and the transported material consists of clastic sediment. Besides transport by water, sediment can in continental environments also be transported by wind or glaciers. Sediment transported by wind is called aeolian and is always very well sorted, while sediment transported by a glacier is called glacial till and is characterized by very poor sorting.

Aeolian deposits can be quite striking. The depositional environment of the Touchet Formation, located in the Northwestern United States, had intervening periods of aridity which resulted in a series of rhythmite layers. Erosional cracks were later infilled with layers of soil material, especially from aeolian processes. The infilled sections formed vertical inclusions in the horizontally deposited layers of the Touchet Formation, and thus provided evidence of the events that intervened in time among the forty-one layers that were deposited.

Sedimentary Facies

Sedimentary environments usually exist alongside each other in certain natural successions. A beach, where sand and gravel is deposited, is usually bounded by a deeper marine environment a little offshore, where finer sediments are deposited at the same time. Behind the beach, there can be dunes (where the dominant deposition is well sorted sand) or a lagoon (where fine clay and organic material is deposited). Every sedimentary environment has its own characteristic deposits. The typical rock formed in a certain environment is called its sedimentary facies. When sedimentary strata accumulate through time, the environment can shift, forming a change in facies in the subsurface at one location. On the other hand, when a rock layer with a certain age is followed laterally, the lithology (the type of rock) and facies eventually change.

Facies can be distinguished in a number of ways: the most common ways are by the lithology (for example: limestone, siltstone or sandstone) or by fossil content. Coral for example only lives in warm and shallow marine environments and fossils of coral are thus typical for shallow marine facies. Facies determined by lithology are called lithofacies; facies determined by fossils are biofacies.

Sedimentary environments can shift their geographical positions through time. Coastlines can shift in the direction of the sea when the sea level drops, when the surface rises due to tectonic forces in the Earth's crust or when a river forms a large delta. In the subsurface, such geographic shifts of sedimentary environments of the past are recorded in shifts in sedimentary facies. This means that sedimentary facies can change either parallel or perpendicular to an imaginary layer of rock with a fixed age, a phenomenon described by Walther's Law.

The situation in which coastlines move in the direction of the continent is called transgression. In the case of transgression, deeper marine facies are deposited over shallower facies, a succession called onlap. Regression is the situation in which a coastline moves in the direction of the sea. With regression, shallower facies are deposited on top of deeper facies, a situation called offlap.

The facies of all rocks of a certain age can be plotted on a map to give an overview of the palaeogeography. A sequence of maps for different ages can give an insight in the development of the regional geography.

Sedimentary Basins

Places where large-scale sedimentation takes place are called sedimentary basins. The amount of sediment that can be deposited in a basin depends on the depth of the basin, the so-called accommodation space. Depth, shape and size of a basin depend on tectonics, movements within the Earth's lithosphere. Where the lithosphere moves upward (tectonic uplift), land eventually rises above sea level, so that and erosion removes material, and the area becomes a source for new sediment. Where the lithosphere moves downward (tectonic subsidence), a basin forms and sedimentation can take place. When the lithosphere keeps subsiding, new accommodation space keeps being created.

A type of basin formed by the moving apart of two pieces of a continent is called a rift basin. Rift basins are elongated, narrow and deep basins. Due to divergent movement, the lithosphere is stretched and thinned, so that the hot asthenosphere rises and heats the overlying rift basin. Apart from continental sediments, rift basins normally also have part of their infill consisting of volcanic deposits. When the basin grows due to continued stretching of the lithosphere, the rift grows and the sea can enter, forming marine deposits.

When a piece of lithosphere that was heated and stretched cools again, its density rises, causing isostatic subsidence. If this subsidence continues long enough the basin is called a sag basin. Examples of sag basins are the regions along passive continental margins, but sag basins can also be found in the interior of continents. In sag basins, the extra weight of the newly deposited sediments is enough to keep the subsidence going in a vicious circle. The total thickness of the sedimentary infill in a sag basins can thus exceed 10 km.

A third type of basin exists along convergent plate boundaries - places where one tectonic plate moves under another into the asthenosphere. The subducting plate bends and forms a fore-arc basin in front of the overriding plate—an elongated, deep asymmetric basin. Fore-arc basins are filled with deep marine deposits and thick sequences of turbidites. Such infill is called flysch. When the convergent movement of the two plates results in continental collision, the basin becomes shallower and develops into a foreland basin. At the same time, tectonic uplift forms a mountain belt in the overriding plate, from which large amounts of material are eroded and transported to the basin. Such erosional material of a growing mountain chain is called molasse and has either a shallow marine or a continental facies.

At the same time, the growing weight of the mountain belt can cause isostatic subsidence in the area of the overriding plate on the other side to the mountain belt. The basin type resulting from this subsidence is called a back-arc basin and is usually filled by shallow marine deposits and molasse.

INFLUENCE OF ASTRONOMICAL CYCLES

In many cases facies changes and other lithological features in sequences of sedimentary rock have a cyclic nature. This cyclic nature was caused by cyclic changes in sediment supply and the sedimentary environment. Most of these cyclic changes are caused by astronomic cycles. Short astronomic cycles can be the difference between the tides or the spring tide every two weeks. On a larger time-scale, cyclic changes in climate and sea level are caused by Milankovitch cycles: cyclic changes in the orientation and/or position of the Earth's rotational axis and orbit around the Sun. There are a number of Milankovitch cycles known, lasting between 10, 000 and 200, 000 years.

Relatively small changes in the orientation of the Earth's axis or length of the seasons can be a major influence on the Earth's climate. An example are the ice ages of the past 2. 6 million years (the Quaternary period), which are assumed to have been caused by astronomic cycles. Climate change can influence the global sea level (and thus the amount of accommodation space in sedimentary basins) and sediment supply from a certain region. Eventually, small changes in astronomic parameters can cause large changes in sedimentary environment and sedimentation.

Sedimentation Rates

The rate at which sediment is deposited differs depending on the location. A channel in a tidal flat can see the deposition of a few metres of sediment in one day, while on the deep ocean floor each year only a few millimetres of sediment accumulate. A distinction can be made between normal sedimentation and sedimentation caused by catastrophic processes. The latter category includes all kinds of sudden exceptional processes like mass movements, rock slides or flooding. Catastrophic processes can see the sudden deposition of a large amount of sediment at once. In some sedimentary environments, most of the total column of sedimentary rock was formed by catastrophic processes, even though the environment is usually a quiet place. Other sedimentary environments are dominated by normal, ongoing sedimentation.

In many cases, sedimentation occurs slowly. In a desert, for example, the wind deposits siliciclastic material (sand or silt) in some spots, or catastrophic flooding of a wadi may cause sudden deposits of large quantities of detrital material, but in most places eolian erosion dominates. The amount of sedimentary rock that forms is not only dependent on the amount of supplied material, but also on how well the material consolidates. Erosion removes most deposited sediment shortly after deposition.

Stratigraphy

That new rock layers are above older rock layers is stated in the principle of superposition. There are usually some gaps in the sequence called unconformities. These represent periods where no new sediments were laid down, or when earlier sedimentary layers raised above sea level and eroded away.

Sedimentary rocks contain important information about the history of the Earth. They contain fossils, the preserved remains of ancient plants and animals. Coal is considered a type of sedimentary rock. The composition of sediments provides us with clues as to the original rock. Differences between successive layers indicate changes to the environment over time. Sedimentary rocks can contain fossils because, unlike most igneous and metamorphic rocks, they form at temperatures and pressures that do not destroy fossil remains.

COMMON METAMORPHIC ROCKS

Metamorphic rocks arise from the transformation of existing rock types, in a process called metamorphism, which means "change in form". The original rock (protolith) is subjected to heat (temperatures greater than 150 to 200 °C) and pressure (1500 bars), causing profound physical and/or chemical change. The protolith may be a sedimentary rock, an igneous rock or another older metamorphic rock.

Metamorphic rocks make up a large part of the Earth's crust and are classified by texture and by chemical and mineral assemblage (metamorphic facies). They may be formed simply by being deep beneath the Earth's surface, subjected to high temperatures and the great pressure of the rock layers above it. They can form from tectonic processes such as continental collisions, which cause horizontal pressure, friction and distortion. They are also formed when rock is heated up by the intrusion of hot molten rock called magma from the Earth's interior. The study of metamorphic rocks (now exposed at the Earth's surface following erosion and uplift) provides information about the temperatures and pressures that occur at great depths within the Earth's crust. Some examples of metamorphic rocks are gneiss, slate, marble, schist, and quartzite.

METAMORPHIC MINERALS

Metamorphic minerals are those that form only at the high temperatures and pressures associated with the process of metamorphism. These minerals, known as index minerals, include sillimanite, kyanite, staurolite, andalusite, and some garnet.

Other minerals, such as olivines, pyroxenes, amphiboles, micas, feldspars, and quartz, may be found in metamorphic rocks, but are not necessarily the result of the process of metamorphism. These minerals formed during the crystallization of igneous rocks. They are stable at high temperatures and pressures and may remain chemically unchanged during the metamorphic process. However, all minerals are stable only within certain limits, and the presence of some minerals in metamorphic rocks indicates the approximate temperatures and pressures at which they formed.

The change in the particle size of the rock during the process of metamorphism is called recrystallization. For instance, the small calcite crystals in the sedimentary rock limestone and chalk change into larger crystals in the metamorphic rock marble, or in metamorphosed sandstone, recrystallization of the original quartz sand grains results in very compact quartzite, also known as metaquartzite, in which the often larger quartz crystals are interlocked. Both high temperatures and pressures contribute to recrystallization. High temperatures allow the atoms and ions in solid crystals to migrate, thus reorganizing the crystals, while high pressures cause solution of the crystals within the rock at their point of contact.

Foliation

The layering within metamorphic rocks is called foliation (derived from the Latin word folia, meaning "leaves"), and it occurs when a rock is being shortened along one axis during recrystallization. This causes the platy or elongated crystals of minerals, such as mica and chlorite, to become rotated such that their long axes are perpendicular to the orientation of shortening. This results in a banded, or foliated rock, with the bands showing the colors of the minerals that formed them.

Textures are separated into foliated and non-foliated categories. Foliated rock is a product of differential stress that deforms the rock in one plane, sometimes creating a plane of cleavage. For example, slate is a foliated metamorphic rock, originating from shale. Non-foliated rock does not have planar patterns of strain.

Rocks that were subjected to uniform pressure from all sides, or those that lack minerals with distinctive growth habits, will not be foliated. Where a rock has been subject to differential stress, the type of foliation that develops depends on the metamorphic grade. For instance, starting with a mudstone, the following sequence develops with increasing temperature: slate is a very fine-grained, foliated metamorphic rock, characteristic of very low grade metamorphism, while phyllite is fine-grained and found in areas of low grade metamorphism, schist is medium to coarse-grained and found in areas of medium grade metamorphism, and gneiss coarse to very coarse-grained, found in areas of high-grade metamorphism. Marble is generally not foliated, which allows its use as a material for sculpture and architecture.

Another important mechanism of metamorphism is that of chemical reactions that occur between minerals without them melting. In the process atoms are exchanged between the minerals, and thus new minerals are formed. Many complex high-temperature reactions may take place, and each mineral assemblage produced provides us with a clue as to the temperatures and pressures at the time of metamorphism.

Metasomatism is the drastic change in the bulk chemical composition of a rock that often occurs during the processes of metamorphism. It is due to the introduction of chemicals from other surrounding rocks. Water may transport these chemicals rapidly over great distances. Because of the role played by water, metamorphic rocks generally contain many elements absent from the original rock, and lack some that originally were present. Still, the introduction of new chemicals is not necessary for recrystallization to occur.

TYPES OF METAMORPHISM

Contact Metamorphism

Contact metamorphism is the name given to the changes that take place when magma is injected into the surrounding solid rock (country rock). The changes that occur are greatest wherever the magma comes into contact with the rock because the temperatures are highest at this boundary and decrease with distance from it. Around the igneous rock that forms from the cooling magma is a metamorphosed zone called a contact metamorphism aureole. Aureoles may show all degrees of metamorphism from the contact area to unmetamorphosed (unchanged) country rock some distance away. The formation of important ore minerals may occur by the process of metasomatism at or near the contact zone.

When a rock is contact altered by an igneous intrusion it very frequently becomes more indurated, and more coarsely crystalline. Many altered rocks of this type were formerly called hornstones, and the term hornfels is often used by geologists to signify those fine grained, compact, non-foliated products of contact metamorphism. A shale may become a dark argillaceous hornfels, full of tiny plates of brownish biotite; a marl or impure limestone may change to a grey, yellow or greenish lime-silicate-hornfels or siliceous marble, tough and splintery, with abundant augite, garnet, wollastonite and other minerals in which calcite is an important component. A diabase or andesite may become a diabase hornfels or andesite hornfels with development of new hornblende and biotite and a partial recrystallization of the original feldspar. Chert or flint may become a finely crystalline quartz rock; sandstones lose their clastic structure and are converted into a mosaic of small close-fitting grains of quartz in a metamorphic rock called quartzite.

If the rock was originally banded or foliated (as, for example, a laminated sandstone or a foliated calc-schist) this character may not be obliterated, and a banded hornfels is the product; fossils even may have their shapes preserved, though entirely recrystallized, and in many contact-altered lavas the vesicles are still visible, though their contents have usually entered into new combinations to form minerals that were not originally present. The minute structures, however, disappear, often completely, if the thermal alteration is very profound. Thus small grains of quartz in a shale are lost or blend with the surrounding particles of clay, and the fine ground-mass of lavas is entirely reconstructed.

By recrystallization in this manner peculiar rocks of very distinct types are often produced. Thus shales may pass into cordierite rocks, or may show large crystals of andalusite (and chiastolite), staurolite, garnet, kyanite and sillimanite, all derived from the aluminous content of the original shale. A considerable amount of mica (both muscovite and biotite) is often simultaneously formed, and the resulting product has a close resemblance to many kinds of schist. Limestones, if pure, are often turned into coarsely crystalline marbles; but if there was an admixture of clay or sand in the original rock such minerals as garnet, epidote, idocrase, wollastonite, will be present. Sandstones when greatly heated may change into coarse quartzites composed of large clear grains of quartz. These more intense stages of alteration are not so commonly seen in igneous rocks, because their minerals, being formed at high temperatures, are not so easily transformed or recrystallized.

In a few cases rocks are fused and in the dark glassy product minute crystals of spinel, sillimanite and cordierite may separate out. Shales are occasionally thus altered by basalt dikes, and feldspathic sandstones may be completely vitrified. Similar changes may be induced in shales by the burning of coal seams or even by an ordinary furnace.

There is also a tendency for metasomatism between the igneous magma and sedimentary country rock, whereby the chemicals in each are exchanged or introduced into the other. Granites may absorb fragments of shale or pieces of basalt. In that case, hybrid rocks called skarn arise, which don't have the characteristics of normal igneous or sedimentary rocks. Sometimes an invading granite magma permeates the rocks around, filling their joints and planes of bedding, etc., with threads of quartz and feldspar. This is very exceptional but instances of it are known and it may take place on a large scale.

Regional metamorphism

Regional metamorphism, also known as dynamic metamorphism, is the name given to changes in great masses of rock over a wide area. Rocks can be metamorphosed simply by being at great depths below the Earth's surface, subjected to high temperatures and the great pressure caused by the immense weight of the rock layers above. Much of the lower continental crust is metamorphic, except for recent igneous intrusions. Horizontal tectonic movements such as the collision of continents create orogenic belts, and cause high temperatures, pressures and deformation in the rocks along these belts. If the metamorphosed rocks are later uplifted and exposed by erosion, they may occur in long belts or other large areas at the surface. The process of metamorphism may have destroyed the original features that could have revealed the rock's previous history. Recrystallization of the rock will destroy the textures and fossils present in sedimentary rocks. Metasomatism will change the original composition.

Regional metamorphism tends to make the rock more indurated and at the same time to give it a foliated, shistose or gneissic texture, consisting of a planar arrangement of the minerals, so that platy or prismatic minerals like mica and hornblende have their longest axes arranged parallel to one another. For that reason many of these rocks split readily in one direction along mica-bearing zones (schists). In gneisses, minerals also tend to be segregated into bands; thus there are seams of quartz and of mica in a mica schist, very thin, but consisting essentially of one mineral. Along the mineral layers composed of soft or fissile minerals the rocks will split most readily, and the freshly split specimens will appear to be faced or coated with this mineral; for example, a piece of mica schist looked at facewise might be supposed to consist entirely of shining scales of mica. On the edge of the specimens, however, the white folia of granular quartz will be visible. In gneisses these alternating folia are sometimes thicker and less regular than in schists, but most importantly less micaceous; they may be lenticular, dying out rapidly. Gneisses also, as a rule, contain more feldspar than schists do, and are tougher and less fissile. Contortion or crumbling of the foliation is by no means uncommon; splitting faces are undulose or puckered. Schistosity and gneissic banding (the two main types of foliation) are formed by directed pressure at elevated temperature, and to interstitial movement, or internal flow arranging the mineral particles while they are crystallizing in that directed pressure field.

Rocks that were originally sedimentary and rocks that were undoubtedly igneous may be metamorphosed into schists and gneisses. If originally of similar composition they may be very difficult to distinguish from one another if the metamorphism has been great. A quartz-porphyry, for example, and a fine feldspathic sandstone, may both be metamorphosed into a grey or pink mica-schist.

Metamorphic Rock Textures

The five basic metamorphic textures with typical rock types are slaty (includes slate and phyllite; the foliation is called "slaty cleavage"), schistose (includes schist; the foliation is called "schistosity"), gneissose (gneiss; the foliation is called "gneissosity"), granoblastic (includes granulite, some marbles and quartzite), and hornfelsic (includes hornfels and skarn).

CHAPTER

STRUCTURAL GEOLOGY 8

Structural geology is the study of the three-dimensional distribution of rock units with respect to their deformational histories. The primary goal of structural geology is to use measurements of present-day rock geometries to uncover information about the history of deformation (strain) in the rocks, and ultimately, to understand the stress field that resulted in the observed strain and geometries. This understanding of the dynamics of the stress field can be linked to important events in the regional geologic past; a common goal is to understand the structural evolution of a particular area with respect to regionally widespread patterns of rock deformation (e. g., mountain building, rifting) due to plate tectonics.

USE AND IMPORTANCE

The study of geologic structures has been of prime importance in economic geology, both petroleum geology and mining geology. Folded and faulted rock strata commonly form traps for the accumulation and concentration of fluids such as petroleum and natural gas. Faulted and structurally complex areas are notable as permeable zones for hydrothermal fluids and the resulting concentration areas for base and precious metal ore deposits. Veins of minerals containing various metals commonly occupy faults and fractures in structurally complex areas. These structurally fractured and faulted zones often occur in association with intrusive igneous rocks. They often also occur around geologic reef complexes and collapse features such as ancient sinkholes. Deposits of gold, silver, copper, lead, zinc, and other metals, are commonly located in structurally complex areas.

Structural geology is a critical part of engineering geology, which is concerned with the physical and mechanical properties of natural rocks. Structural fabrics and defects such as faults, folds, foliations and joints are internal weaknesses of rocks which may affect the stability of human engineered structures such as dams, road cuts, open pit mines and underground mines or road tunnels.

Geotechnical risk, including earthquake risk can only be investigated by inspecting a combination of structural geology and geomorphology. In addition areas of karst landscapes which are underlain by underground caverns and potential sinkholes or collapse features are of importance for these scientists. In addition, areas of steep slopes are potential collapse or landslide hazards.

Environmental geologists and hydrogeologists or hydrologists need to understand structural geology because structures are sites of groundwater flow and penetration, which may affect, for instance, seepage of toxic substances from waste dumps, or seepage of salty water into aquifers.

Plate tectonics is a theory developed during the 1960s which describes the movement of continents by way of the separation and collision of crustal plates. It is in a sense structural geology on a planet scale, and is used throughout structural geology as a framework to analyze and understand global, regional, and local scale features.

Methods

Structural geologists use a variety of methods to (first) measure rock geometries, (second) reconstruct their deformational histories, and (third) calculate the stress field that resulted in that deformation.

Geometries

Primary data sets for structural geology are collected in the field. Structural geologists measure a variety of planar features (bedding planes, foliation planes, fold axial planes, fault planes, and joints), and linear features (stretching lineations, in which minerals are ductily extended; fold axes; and intersection lineations, the trace of a planar feature on another planar surface).

Measurement Conventions

The inclination of a planar structure in geology is measured by strike and dip. The strike is the line of intersection between the planar feature and a horizontal plane, taken according to the right hand convention, and the dip is the magnitude of the inclination, below horizontal, at right angles to strike. For example; striking 25 degrees East of North, dipping 45 degrees Southeast, recorded as N25E, 45SE.

Alternatively, dip and dip direction may be used as this is absolute. Dip direction is measured in 360 degrees, generally clockwise from North. For example, a dip of 45 degrees towards 115 degrees azimuth, recorded as 45/115.

The term hade is occasionally used and is the deviation of a plane from vertical i. e. (90°-dip).

Fold axis plunge is measured in dip and dip direction (strictly, plunge and azimuth of plunge). The orientation of a fold axial plane is measured in strike and dip or dip and dip direction.

Lineations are measured in terms of dip and dip direction, if possible. Often lineations occur expressed on a planar surface and can be difficult to measure directly. In this case, the lineation may be measured from the horizontal as a rake or pitch upon the surface.

Rake is measured by placing a protractor flat on the planar surface, with the flat edge horizontal and measuring the angle of the lineation clockwise from horizontal. The orientation of the lineation can then be calculated from the rake and strike-dip information of the plane it was measured from, using a stereographic projection.

If a fault has lineations formed by movement on the plane, e. g. ; slickensides, this is recorded as a lineation, with a rake, and annotated as to the indication of throw on the fault.

Generally it is easier to record strike and dip information of planar structures in dip/dip direction format as this will match all the other structural information you may be recording about folds, lineations, etc., although there is an advantage to using different formats that discriminate between planar and linear data.

Plane, fabric, fold and deformation conventions

The convention for analysing structural geology is to identify the planar structures, often called planar fabrics because this implies a textural formation, the linear structures and, from analysis of these, unravel deformations.

Planar structures are named according to their order of formation, with original sedimentary layering the lowest at S0. Often it is impossible to identify S0 in highly deformed rocks, so numbering may be started at an arbitrary number or given a letter (SA, for instance). In cases where there is a bedding-plane foliation caused by burial metamorphism or diagenesis this may be enumerated as S0a.

If there are folds, these are numbered as F1, F2, etc. Generally the axial plane foliation or cleavage of a fold is created during folding, and the number convention should match. For example, an F2 fold should have an S2 axial foliation.

Deformations are numbered according to their order of formation with the letter D denoting a deformation event. For example D1, D2, D3. Folds and foliations, because they are formed by deformation events, should correlate with these events. For example an F2 fold, with an S2 axial plane foliation would be the result of a D2 deformation.

Metamorphic events may span multiple deformations. Sometimes it is useful to identify them similarly to the structural features for which they are responsible, e. g. ; M2. This may be possible by observing porphyroblast formation in cleavages of known deformation age, by identifying metamorphic mineral assemblages created by different events, or via geochronology.

Intersection lineations in rocks, as they are the product of the intersection of two planar structures, are named according to the two planar structures from which they are formed. For instance, the intersection lineation of a S1 cleavage and bedding is the L1-0 intersection lineation (also known as the cleavage-bedding lineation).

Stretching lineations may be difficult to quantify, especially in highly stretched ductile rocks where minimal foliation information is preserved. Where possible, when correlated with deformations (as few are formed in folds, and many are not strictly associated with planar foliations), they may be identified similar to planar surfaces and folds, e. g. ; L1, L2. For convenience some geologists prefer to annotate them with a subscript S, for example Ls1 to differentiate them from intersection lineations, though this is generally redundant.

STEREOGRAPHIC PROJECTIONS

Stereographic projection of structural strike and dip measurements is a powerful method for analyzing the nature and orientation of deformation stresses, lithological units and penetrative fabrics.

Rock macro-structures

On a large scale, structural geology is the study of the three dimensional relationships of stratigraphic units to one another within terranes of rock or within geological regions.

This branch of structural geology deals mainly with the orientation, deformation and relationships of stratigraphy (bedding), which may have been faulted, folded or given a foliation by some tectonic event. This is mainly a geometric science, from which cross sections and three dimensional block models of rocks, regions, terranes and parts of the Earth's crust can be generated.

Study of regional structure is important in understanding orogeny, plate tectonics and more specifically in the oil, gas and mineral exploration industries as structures such as faults, folds and unconformities are primary controls on ore mineralisation and oil traps.

Modern regional structure is being investigated using seismic tomography and seismic reflection in three dimensions, providing unrivaled images of the Earth's interior, its faults and the deep crust. Further information from geophysics such as gravity and airborne magnetics can provide information on the nature of rocks imaged in the deep crust.

Rock microstructures

Rock microstructure or texture of rocks is studied by structural geologists on a small scale to provide detailed information mainly about metamorphic rocks and some features of sedimentary rocks, most often if they have been folded.

Textural study involves measurement and characterisation of foliations, crenulations, metamorphic minerals, and timing relationships between these structural features and mineralogical features.

Usually this involves collection of hand specimens, which may be cut to provide petrographic thin sections which are analysed under a petrographic microscope.

Microstructural analysis finds application also in multi-scale statistical analysis, aimed to analyze some rock features showing scale invariance.

Kinematics

Geologists use their measurements of rock geometries to understand histories of strain in the rocks. Strain can take the form of brittle faulting and ductile folding and shearing. Brittle deformation takes place in the shallow crust, and ductile deformation takes place in the deeper crust, where temperatures and pressures are higher.

Stress Fields

By understanding the constitutive relationships between stress and strain in rocks, geologists can translate the observed patterns of rock deformation into a stress field during the geologic past. The following list of features are typically used to determine stress fields from deformational structures.

- In perfectly brittle rocks, faulting occurs at 30° to the greatest compressional stress. (Byerlee's Law)
- The greatest compressive stress is normal to fold axial planes.

GROUNDWATER MOVEMENT 9

Groundwater is the water located beneath Earth's surface in soil pore spaces and in the fractures of rock formations. A unit of rock or an unconsolidated deposit is called an aquifer when it can yield a usable quantity of water. The depth at which soil pore spaces or fractures and voids in rock become completely saturated with water is called the water table. Groundwater is recharged from, and eventually flows to, the surface naturally; natural discharge often occurs at springs and seeps, and can form oases or wetlands. Groundwater is also often withdrawn for agricultural, municipal, and industrial use by constructing and operating extraction wells. The study of the distribution and movement of groundwater is hydrogeology, also called groundwater hydrology.

Typically, groundwater is thought of as liquid water flowing through shallow aquifers, but, in the technical sense, it can also include soil moisture, permafrost (frozen soil), immobile water in very low permeability bedrock, and deep geothermal or oil formation water. Groundwater is hypothesized to provide lubrication that can possibly influence the movement of faults. It is likely that much of Earth's subsurface contains some water, which may be mixed with other fluids in some instances. Groundwater may not be confined only to Earth. The formation of some of the landforms observed on Mars may have been influenced by groundwater. There is also evidence that liquid water may also exist in the subsurface of Jupiter's moon Europa.

AQUIFERS

An aquifer is a layer of porous substrate that contains and transmits groundwater. When water can flow directly between the surface and the saturated zone of an aquifer, the aquifer is unconfined. The deeper parts of unconfined aquifers are usually more saturated since gravity causes water to flow downward.

The upper level of this saturated layer of an unconfined aquifer is called the water table or phreatic surface. Below the water table, where in general all pore spaces are saturated with water, is the phreatic zone.

Substrate with low porosity that permits limited transmission of groundwater is known as an aquitard. An aquiclude is a substrate with porosity that is so low it is virtually impermeable to groundwater.

A confined aquifer is an aquifer that is overlain by a relatively impermeable layer of rock or substrate such as an aquiclude or aquitard. If a confined aquifer follows a downward grade from its recharge zone, groundwater can become pressurized as it flows. This can create artesian wells that flow freely without the need of a pump and rise to a higher elevation than the static water table at the above, unconfined, aquifer.

The characteristics of aquifers vary with the geology and structure of the substrate and topography in which they occur. In general, the more productive aquifers occur in sedimentary geologic formations. By comparison, weathered and fractured crystalline rocks yield smaller quantities of groundwater in many environments. Unconsolidated to poorly cemented alluvial materials that have accumulated as valley-filling sediments in major river valleys and geologically subsiding structural basins are included among the most productive sources of groundwater.

The high specific heat capacity of water and the insulating effect of soil and rock can mitigate the effects of climate and maintain groundwater at a relatively steady temperature. In some places where groundwater temperatures are maintained by this effect at about 10 °C (50 °F), groundwater can be used for controlling the temperature inside structures at the surface. For example, during hot weather relatively cool groundwater can be pumped through radiators in a home and then returned to the ground in another well. During cold seasons, because it is relatively warm, the water can be used in the same way as a source of heat for heat pumps that is much more efficient than using air.

Groundwater is often cheaper, more convenient and less vulnerable to pollution than surface water. Therefore, it is commonly used for public water supplies. Groundwater provides the largest source of usable water storage in the United States. Underground reservoirs contain far more water than the capacity of all surface reservoirs and lakes, including the Great Lakes. Many municipal water supplies are derived solely from groundwater.

The volume of groundwater in an aquifer can be estimated by measuring water levels in local wells and by examining geologic records from well-drilling to determine the extent, depth and thickness of water-bearing sediments and rocks. Before an investment is made in production wells, test wells may be drilled to measure the depths at which water is encountered and collect samples of soils, rock and water for laboratory analyses. Pumping tests can be performed in test wells to determine flow characteristics of the aquifer.

Polluted ground water is less visible, but more difficult to clean up, than pollution in rivers and lakes. Ground water pollution most often results from improper disposal of wastes on land. Major sources include industrial and household chemicals and garbage landfills, industrial waste lagoons, tailings and process wastewater from mines, oil field brine pits, leaking underground oil storage tanks and pipelines, sewage sludge and septic systems. Polluted groundwater is mapped by sampling soils and groundwater near suspected or known sources of pollution, to determine the extent of the pollution, and to aid in the design of groundwater remediation systems. Preventing groundwater pollution near potential sources such as landfills requires lining the bottom of a landfill with watertight materials, collecting any leachate with drains, and keeping rainwater off any potential contaminants, along with regular monitoring of nearby groundwater to verify that contaminants have not leaked into the groundwater.

The danger of pollution of municipal supplies is minimized by locating wells in areas of deep ground water and impermeable soils, and careful testing and monitoring of the aquifer and nearby potential pollution sources.

Water cycle

Groundwater makes up about twenty percent of the world's fresh water supply, which is about 0. 61% of the entire world's water, including oceans and permanent ice. Global groundwater storage is roughly equal to the total amount of freshwater stored in the snow and ice pack, including the north and south poles. This makes it an important resource that can act as a natural storage that can buffer against shortages of surface water, as in during times of drought.

Groundwater is naturally replenished by surface water from precipitation, streams, and rivers when this recharge reaches the water table.

Groundwater can be a long-term 'reservoir' of the natural water cycle (with residence times from days to millennia), as opposed to short-term water reservoirs like the atmosphere and fresh surface water (which have residence times from minutes to years). The figure shows how deep groundwater (which is quite distant from the surface recharge) can take a very long time to complete its natural cycle.

The Great Artesian Basin in central and eastern Australia is one of the largest confined aquifer systems in the world, extending for almost 2 million km2. By analysing the trace elements in water sourced from deep underground, hydrogeologists have been able to determine that water extracted from these aquifers can be more than 1 million years old.

By comparing the age of groundwater obtained from different parts of the Great Artesian Basin, hydrogeologists have found it increases in age across the basin. Where water recharges the aquifers along the Eastern Divide, ages are young. As groundwater flows westward across the continent, it increases in age, with the oldest groundwater occurring in the western parts. This means that in order to have travelled almost 1000 km from the source of recharge in 1 million years, the groundwater flowing through the Great Artesian Basin travels at an average rate of about 1 metre per year.

Recent research has demonstrated that evaporation of groundwater can play a significant role in the local water cycle, especially in arid regions. Scientists in Saudi Arabia have proposed plans to recapture and recycle this evaporative moisture for crop irrigation. In the opposite photo, a 50-centimeter-square reflective carpet, made of small adjacent plastic cones, was placed in a plant-free dry desert area for five months, without rain or irrigation. It managed to capture and condense enough ground vapor to bring to life naturally buried seeds underneath it, with a green area of about 10% of the carpet area. It is expected that, if seeds were put down before placing this carpet, a much wider area would become green.

ISSUES

Overview

Certain problems have beset the use of groundwater around the world. Just as river waters have been over-used and polluted in many parts of the world, so too have aquifers. The big difference is that aquifers are out of sight. The other major problem is that water management agencies, when calculating the "sustainable yield" of aquifer and river water, have often counted the same water twice, once in the aquifer, and once in its connected river. This problem, although understood for centuries, has persisted, partly through inertia within government agencies. In Australia, for example, prior to the statutory reforms initiated by the Council of Australian Governments water reform framework in the 1990s, many Australian states managed groundwater and surface water through separate government agencies, an approach beset by rivalry and poor communication.

In general, the time lags inherent in the dynamic response of groundwater to development have been ignored by water management agencies, decades after scientific understanding of the issue was consolidated. In brief, the effects of groundwater overdraft (although undeniably real) may take decades or centuries to manifest themselves. In a classic study in 1982, Bredehoeft and colleagues modeled a situation where groundwater extraction in an intermontane basin withdrew the entire annual recharge, leaving 'nothing' for the natural groundwater-dependent vegetation community. Even when the borefield was situated close to the vegetation, 30% of the original vegetation demand could still be met by the lag inherent in the system after 100 years. By year 500, this had reduced to 0%, signalling complete death of the groundwater-dependent vegetation. The science has been available to make these calculations for decades; however, in general water management agencies have ignored effects that will appear outside the rough timeframe of political elections (3 to 5 years). Marios Sophocleous argued strongly that management agencies must define and use appropriate timeframes in groundwater planning. This will mean calculating groundwater withdrawal permits based on predicted effects decades, sometimes centuries in the future.

As water moves through the landscape, it collects soluble salts, mainly sodium chloride. Where such water enters the atmosphere through evapotranspiration, these salts are left behind. In irrigation districts, poor drainage of soils and surface aquifers can result in water tables' coming to the surface in low-lying areas. Major land degradation problems of soil salinity and waterlogging result, combined with increasing levels of salt in surface waters. As a consequence, major damage has occurred to local economies and environments.

Four important effects are worthy of brief mention. First, flood mitigation schemes, intended to protect infrastructure built on floodplains, have had the unintended consequence of reducing aquifer recharge associated with natural flooding. Second, prolonged depletion of groundwater in extensive aquifers can result in land subsidence, with associated infrastructure damage – as well as, third, saline intrusion. Fourth, draining acid sulphate soils, often found in low-lying coastal plains, can result in acidification and pollution of formerly freshwater and estuarine streams.

Another cause for concern is that groundwater drawdown from over-allocated aquifers has the potential to cause severe damage to both terrestrial and aquatic ecosystems – in some cases very conspicuously but in others quite imperceptibly because of the extended period over which the damage occurs.

Overdraft

Groundwater is a highly useful and often abundant resource. However, over-use, or overdraft, can cause major problems to human users and to the environment. The most evident problem (as far as human groundwater use is concerned) is a lowering of the water table beyond the reach of existing wells. As a consequence, wells must be drilled deeper to reach the groundwater; in some places (e. g., California, Texas, and India) the water table has dropped hundreds of feet because of extensive well pumping. In the Punjab region of India, for example, groundwater levels have dropped 10 meters since 1979, and the rate of depletion is accelerating. A lowered water table may, in turn, cause other problems such as groundwater-related subsidence and saltwater intrusion.

Groundwater is also ecologically important. The importance of groundwater to ecosystems is often overlooked, even by freshwater biologists and ecologists. Groundwaters sustain rivers, wetlands, and lakes, as well as subterranean ecosystems within karst or alluvial aquifers.

Not all ecosystems need groundwater, of course. Some terrestrial ecosystems – for example, those of the open deserts and similar arid environments – exist on irregular rainfall and the moisture it delivers to the soil, supplemented by moisture in the air. While there are other terrestrial ecosystems in more hospitable environments where groundwater plays no central role, groundwater is in fact fundamental to many of the world's major ecosystems. Water flows between groundwaters and surface waters. Most rivers, lakes, and wetlands are fed by, and (at other places or times) feed groundwater, to varying degrees. Groundwater feeds soil moisture through percolation, and many terrestrial vegetation communities depend directly on either groundwater or the percolated soil moisture above the aquifer for at least part of each year. Hyporheic zones (the mixing zone of streamwater and groundwater) and riparian zones are examples of ecotones largely or totally dependent on groundwater.

Subsidence

Subsidence occurs when too much water is pumped out from underground, deflating the space below the above-surface, and thus causing the ground to collapse. The result can look like craters on plots of land. This occurs because, in its natural equilibrium state, the hydraulic pressure of groundwater in the pore spaces of the aquifer and the aquitard supports some of the weight of the overlying sediments. When groundwater is removed from aquifers by excessive pumping, pore pressures in the aquifer drop and compression of the aquifer may occur. This compression may be partially recoverable if pressures rebound, but much of it is not. When the aquifer gets compressed, it may cause land subsidence, a drop in the ground surface. The city of New Orleans, Louisiana is actually below sea level today, and its subsidence is partly caused by removal of groundwater from the various aquifer/aquitard systems beneath it. In the first half of the 20th century, the city of San Jose, California dropped 13 feet from land subsidence caused by overpumping; this subsidence has been halted with improved groundwater management.

Seawater intrusion

In general, in very humid or undeveloped regions, the shape of the water table mimics the slope of the surface. The recharge zone of an aquifer near the seacoast is likely to be inland, often at considerable distance. In these coastal areas, a lowered water table may induce sea water to reverse the flow toward the land. Sea water moving inland is called a saltwater intrusion. In alternative fashion, salt from mineral beds may leach into the groundwater of its own accord.

Pollution

Groundwater pollution, from pollutants released to the ground that can work their way down into groundwater, can create a contaminant plume within an aquifer.

Pollution can occur from landfills, naturally occurring arsenic, on-site sanitation systems or other point sources, such as petrol stations or leaking sewers.

Movement of water and dispersion within the aquifer spreads the pollutant over a wider area, its advancing boundary often called a plume edge, which can then intersect with groundwater wells or daylight into surface water such as seeps and springs, making the water supplies unsafe for humans and wildlife. Different mechanism have influence on the transport of pollutants, e. g. diffusion, adsorption, precipitation, decay, in the groundwater. The interaction of groundwater contamination with surface waters is analyzed by use of hydrology transport models.

Government Regulations

In the United States, laws regarding ownership and use of groundwater are generally state laws; however, regulation of groundwater to minimize pollution of groundwater is by both states and the federal-level Environmental Protection Agency. Ownership and use rights to groundwater typically follow one of three main systems:

RULE OF CAPTURE

The Rule of Capture provides each landowner the ability to capture as much groundwater as they can put to a beneficial use, but they are not guaranteed any set amount of water. As a result, well-owners are not liable to other landowners for taking water from beneath their land. State laws or regulations will often define "beneficial use", and sometimes place other limits, such as disallowing groundwater extraction which causes subsidence on neighboring property.

Riparian Rights

Limited private ownership rights similar to riparian rights in a surface stream. The amount of groundwater right is based on the size of the surface area where each landowner gets a corresponding amount of the available water. Once adjudicated, the maximum amount of the water right is set, but the right can be decreased if the total amount of available water decreases as is likely during a drought. Landowners may sue others for encroaching upon their groundwater rights, and water pumped for use on the overlying land takes preference over water pumped for use off the land.

Environmental Protection of Groundwater

In November 2006, the Environmental Protection Agency published the Ground Water Rule in the United States Federal Register. The EPA was worried that the ground water system would be vulnerable to contamination from fecal matter. The point of the rule was to keep microbial pathogens out of public water sources. The 2006 Ground Water Rule was an amendment of the 1996 Safe Drinking Water Act.

Others

Reasonable Use Rule (American Rule)

This rule does not guarantee the landowner a set amount of water, but allows unlimited extraction as long as the result does not unreasonably damage other wells or the aquifer system. Usually this rule gives great weight to historical uses and prevents new uses that interfere with the prior use.

Groundwater scrutiny upon real estate property transactions in the US

In the US, upon commercial real estate property transactions both groundwater and soil are the subjects of scrutiny, with a Phase I Environmental Site Assessment normally being prepared to investigate and disclose potential pollution issues. In the San Fernando Valley of California, real estate contracts for property transfer below the Santa Susana Field Laboratory (SSFL) and eastward have clauses releasing the seller from liability for groundwater contamination consequences from existing or future pollution of the Valley Aquifer.

CHAPTER

STRATIGRAPHY 10

Stratigraphy is a branch of geology which studies rock layers (strata) and layering (stratification). It is primarily used in the study of sedimentary and layered volcanic rocks. Stratigraphy includes two related subfields: lithologic stratigraphy or lithostratigraphy, and biologic stratigraphy or biostratigraphy.

HISTORICAL DEVELOPMENT

Nicholas Steno established the theoretical basis for stratigraphy when he introduced the law of superposition, the principle of original horizontality and the principle of lateral continuity in a 1669 work on the fossilization of organic remains in layers of sediment.

The first practical large-scale application of stratigraphy was by William Smith in the 1790s and early 1800s. Smith, known as the "Father of English geology", created the first geologic map of England and first recognized the significance of strata or rock layering and the importance of fossil markers for correlating strata. Another influential application of stratigraphy in the early 1800s was a study by Georges Cuvier and Alexandre Brongniart of the geology of the region around Paris.

Lithostratigraphy

Lithostratigraphy, or lithologic stratigraphy, provides the most obvious visible layering. It deals with the physical contrasts in lithology, or rock type. Such layers can occur both vertically – in layering or bedding of varying rock types – and laterally – reflecting changing environments of deposition (known as facies change). Key concepts in stratigraphy involve understanding how certain geometric relationships between rock layers arise and what these geometries mean in terms of the depositional environment. Stratigraphers have codified a basic concept of their discipline in the law of superposition, which simply states that, in an undeformed stratigraphic sequence, the oldest strata occur at the base of the sequence.

Chemostratigraphy studies the changes in the relative proportions of trace elements and isotopes within and between lithologic units. Carbon and oxygen isotope ratios vary with time, and researchers can use them to map subtle changes that occurred in the paleoenvironment. This has led to the specialized field of isotopic stratigraphy.

Cyclostratigraphy documents the often cyclic changes in the relative proportions of minerals (particularly carbonates), grain size, or thickness of sediment layers (varves) and of fossil diversity with time, related to seasonal or longer term changes in palaeoclimates.

Biostratigraphy

Biostratigraphy or paleontologic stratigraphy is based on fossil evidence in the rock layers. Strata from widespread locations containing the same fossil fauna and flora are correlatable in time. Biologic stratigraphy was based on William Smith's principle of faunal succession, which predated, and was one of the first and most powerful lines of evidence for, biological evolution. It provides strong evidence for the formation (speciation) and extinction of species. The geologic time scale was developed during the 19th century, based on the evidence of biologic stratigraphy and faunal succession. This timescale remained a relative scale until the development of radiometric dating, which gave it and the stratigraphy it was based on an absolute time framework, leading to the development of chronostratigraphy.

One important development is the Vail curve, which attempts to define a global historical sea-level curve according to inferences from worldwide stratigraphic patterns. Stratigraphy is also commonly used to delineate the nature and extent of hydrocarbon-bearing reservoir rocks, seals, and traps in petroleum geology.

Chronostratigraphy

Chronostratigraphy is the branch of stratigraphy that studies the absolute, not relative, age of rock strata. The branch is concerned with deriving geochronological data for rock units, both directly and inferentially, so that a sequence of time-relative events of rocks within a region can be derived. In essence, chronostratigraphy seeks to understand the geologic history of rocks and regions.

The ultimate aim of chronostratigraphy is to arrange the sequence of deposition and the time of deposition of all rocks within a geological region and, eventually, the entire geologic record of the earth.

A gap or missing strata in the geological record of an area is called a stratigraphic hiatus. This may be the result of lack of sediment deposition or it may be due to removal by erosion, in which case it may be called a vacuity. It is called a hiatus because deposition was on hold for a period of time. A physical gap may represent both a period of non-deposition and a period of erosion. A fault may cause the appearance of a hiatus.

MAGNETOSTRATIGRAPHY

Magnetostratigraphy is a chronostratigraphic technique used to date sedimentary and volcanic sequences. The method works by collecting oriented samples at measured intervals throughout a section. The samples are analyzed to determine their detrital remanent magnetism (DRM), that is, the polarity of Earth's magnetic field at the time a stratum was deposited. For sedimentary rocks, this is possible because, when very fine-grained magnetic minerals (< 17 ìm) fall through the water column, they orient themselves with Earth's magnetic field. Upon burial, that orientation is preserved. The minerals behave like tiny compasses. For volcanic rocks, magnetic minerals, which form in the melt, are fixed in place upon crystallization or freezing of the lava and are oriented with the ambient magnetic field.

Oriented paleomagnetic core samples are collected in the field; mudstones, siltstones, and very fine-grained sandstones are the preferred lithologies because the magnetic grains are finer and more likely to orient with the ambient field during deposition. If the ancient magnetic field were oriented similar to today's field (North Magnetic Pole near the North Rotational Pole), the strata would retain a normal polarity. If the data indicate that the North Magnetic Pole were near the South Rotational Pole, the strata would exhibit reversed polarity.

Results of the individual samples are analyzed by removing the natural remanent magnetization (NRM) to reveal the DRM. Following statistical analysis, the results are used to generate a local magnetostratigraphic column that can then be compared against the Global Magnetic Polarity Time Scale.

This technique is used to date sequences that generally lack fossils or interbedded igneous rocks. The continuous nature of the sampling means that it is also a powerful technique for the estimation of sediment-accumulation rates.

Archaeological stratigraphy

In the field of archaeology, soil stratigraphy is used to better understand the processes that form and protect archaeological sites. Since the law of superposition holds true, it can help date finds or features from each context; these finds and features can be placed in sequence and the dates interpolated. Phases of activity can also often be seen through stratigraphy, especially when a trench or feature is viewed in section (profile). Because pits and other features can be dug down into earlier levels, not all material at the same absolute depth is necessarily of the same age; close attention has to be paid to the archeological layers. The Harris-matrix is a tool to depict complex stratigraphic relations when they are found, for example, in the context of urban archaeology.

EARTHQUAKE

An earthquake (also known as a quake, tremor or temblor) is the perceptible shaking of the surface of the Earth, which can be violent enough to destroy major buildings and kill thousands of people. The severity of the shaking can range from barely felt to violent enough to toss people around. Earthquakes have destroyed whole cities. They result from the sudden release of energy in the Earth's crust that creates seismic waves. The seismicity, seismism or seismic activity of an area refers to the frequency, type and size of earthquakes experienced over a period of time.

Earthquakes are measured using observations from seismometers. The moment magnitude is the most common scale on which earthquakes larger than approximately 5 are reported for the entire globe. The more numerous earthquakes smaller than magnitude 5 reported by national seismological observatories are measured mostly on the local magnitude scale, also referred to as the Richter magnitude scale. These two scales are numerically similar over their range of validity. Magnitude 3 or lower earthquakes are mostly almost imperceptible or weak and magnitude 7 and over potentially cause serious damage over larger areas, depending on their depth. The largest earthquakes in historic times have been of magnitude slightly over 9, although there is no limit to the possible magnitude. The most recent large earthquake of magnitude 9.0 or larger was a 9.0 magnitude earthquake in Japan in 2011 (as of March 2014), and it was the largest Japanese earthquake since records began. Intensity of shaking is measured on the modified Mercalli scale. The shallower an earthquake, the more damage to structures it causes, all else being equal.

At the Earth's surface, earthquakes manifest themselves by shaking and sometimes displacement of the ground. When the epicenter of a large earthquake is located offshore, the seabed may be displaced sufficiently to cause a tsunami. Earthquakes can also trigger landslides, and occasionally volcanic activity.

In its most general sense, the word earthquake is used to describe any seismic event — whether natural or caused by humans — that generates seismic waves. Earthquakes are caused mostly by rupture of geological faults, but also by other events such as volcanic activity, landslides, mine blasts, and nuclear tests. An earthquake's point of initial rupture is called its focus or hypocenter. The epicenter is the point at ground level directly above the hypocenter.

NATURALLY OCCURRING EARTHQUAKES

Tectonic earthquakes occur anywhere in the earth where there is sufficient stored elastic strain energy to drive fracture propagation along a fault plane. The sides of a fault move past each other smoothly and aseismically only if there are no irregularities or asperities along the fault surface that increase the frictional resistance. Most fault surfaces do have such asperities and this leads to a form of stick-slip behaviour. Once the fault has locked, continued relative motion between the plates leads to increasing stress and therefore, stored strain energy in the volume around the fault surface. This continues until the stress has risen sufficiently to break through the asperity, suddenly allowing sliding over the locked portion of the fault, releasing the stored energy. This energy is released as a combination of radiated elastic strain seismic waves, frictional heating of the fault surface, and cracking of the rock, thus causing an earthquake. This process of gradual build-up of strain and stress punctuated by occasional sudden earthquake failure is referred to as the elastic-rebound theory. It is estimated that only 10 percent or less of an earthquake's total energy is radiated as seismic energy. Most of the earthquake's energy is used to power the earthquake fracture growth or is converted into heat generated by friction. Therefore, earthquakes lower the Earth's available elastic potential energy and raise its temperature, though these changes are negligible compared to the conductive and convective flow of heat out from the Earth's deep interior.

Earthquake Fault Types

There are three main types of fault, all of which may cause an interplate earthquake: normal, reverse (thrust) and strike-slip. Normal and reverse faulting are examples of dip-slip, where the displacement along the fault is in the direction of dip and movement on them involves a vertical component. Normal faults occur mainly in areas where the crust is being extended such as a divergent boundary. Reverse faults occur in areas where the crust is being shortened such as at a convergent boundary. Strike-slip faults are steep structures where the two sides of the fault slip horizontally past each other; transform boundaries are a particular type of strike-slip fault. Many earthquakes are caused by movement on faults that have components of both dip-slip and strike-slip; this is known as oblique slip.

Reverse faults, particularly those along convergent plate boundaries are associated with the most powerful earthquakes, megathrust earthquakes, including almost all of those of magnitude 8 or more. Strike-slip faults, particularly continental transforms, can produce major earthquakes up to about magnitude 8. Earthquakes associated with normal faults are generally less than magnitude 7. For every unit increase in magnitude, there is a roughly thirtyfold increase in the energy released. For instance, an earthquake of magnitude 6. 0 releases approximately 30 times more energy than a 5. 0 magnitude earthquake and a 7. 0 magnitude earthquake releases 900 times (30 × 30) more energy than a 5. 0 magnitude of earthquake. An 8. 6 magnitude earthquake releases the same amount of energy as 10, 000 atomic bombs that were used in World War II.

This is so because the energy released in an earthquake, and thus its magnitude, is proportional to the area of the fault that ruptures and the stress drop. Therefore, the longer the length and the wider the width of the faulted area, the larger the resulting magnitude. The topmost, brittle part of the Earth's crust, and the cool slabs of the tectonic plates that are descending down into the hot mantle, are the only parts of our planet which can store elastic energy and release it in fault ruptures. Rocks hotter than about 300 degrees Celsius flow in response to stress; they do not rupture in earthquakes. The maximum observed lengths of ruptures and mapped faults (which may break in a single rupture) are approximately 1000 km. Examples are the earthquakes in Chile, 1960; Alaska, 1957; Sumatra, 2004, all in subduction zones. The longest earthquake ruptures on strike-slip faults, like the San Andreas Fault (1857, 1906), the North Anatolian Fault in Turkey (1939) and the Denali Fault in Alaska (2002), are about half to one third as long as the lengths along subducting plate margins, and those along normal faults are even shorter.

The most important parameter controlling the maximum earthquake magnitude on a fault is however not the maximum available length, but the available width because the latter varies by a factor of 20. Along converging plate margins, the dip angle of the rupture plane is very shallow, typically about 10 degrees. Thus the width of the plane within the top brittle crust of the Earth can become 50 to 100 km (Japan, 2011; Alaska, 1964), making the most powerful earthquakes possible.

Strike-slip faults tend to be oriented near vertically, resulting in an approximate width of 10 km within the brittle crust, thus earthquakes with magnitudes much larger than 8 are not possible. Maximum magnitudes along many normal faults are even more limited because many of them are located along spreading centers, as in Iceland, where the thickness of the brittle layer is only about 6 km.

In addition, there exists a hierarchy of stress level in the three fault types. Thrust faults are generated by the highest, strike slip by intermediate, and normal faults by the lowest stress levels. This can easily be understood by considering the direction of the greatest principal stress, the direction of the force that 'pushes' the rock mass during the faulting. In the case of normal faults, the rock mass is pushed down in a vertical direction, thus the pushing force (greatest principal stress) equals the weight of the rock mass itself. In the case of thrusting, the rock mass 'escapes' in the direction of the least principal stress, namely upward, lifting the rock mass up, thus the overburden equals the least principal stress. Strike-slip faulting is intermediate between the other two types described above. This difference in stress regime in the three faulting environments can contribute to differences in stress drop during faulting, which contributes to differences in the radiated energy, regardless of fault dimensions.

Earthquakes away from plate boundaries

Where plate boundaries occur within the continental lithosphere, deformation is spread out over a much larger area than the plate boundary itself. In the case of the San Andreas fault continental transform, many earthquakes occur away from the plate boundary and are related to strains developed within the broader zone of deformation caused by major irregularities in the fault trace (e. g., the "Big bend" region). The Northridge earthquake was associated with movement on a blind thrust within such a zone. Another example is the strongly oblique convergent plate boundary between the Arabian and Eurasian plates where it runs through the northwestern part of the Zagros mountains. The deformation associated with this plate boundary is partitioned into nearly pure thrust sense movements perpendicular to the boundary over a wide zone to the southwest and nearly pure strike-slip motion along the Main Recent Fault close to the actual plate boundary itself. This is demonstrated by earthquake focal mechanisms.

All tectonic plates have internal stress fields caused by their interactions with neighbouring plates and sedimentary loading or unloading (e. g. deglaciation). These stresses may be sufficient to cause failure along existing fault planes, giving rise to intraplate earthquakes.

Shallow-focus and deep-focus earthquakes

The majority of tectonic earthquakes originate at the ring of fire in depths not exceeding tens of kilometers. Earthquakes occurring at a depth of less than 70 km are classified as 'shallow-focus' earthquakes, while those with a focal-depth between 70 and 300 km are commonly termed 'mid-focus' or 'intermediate-depth' earthquakes. In subduction zones, where older and colder oceanic crust descends beneath another tectonic plate, deep-focus earthquakes may occur at much greater depths (ranging from 300 up to 700 kilometers). These seismically active areas of subduction are known as Wadati-Benioff zones. Deep-focus earthquakes occur at a depth where the subducted lithosphere should no longer be brittle, due to the high temperature and pressure. A possible mechanism for the generation of deep-focus earthquakes is faulting caused by olivine undergoing a phase transition into a spinel structure.

Earthquakes and volcanic activity

Earthquakes often occur in volcanic regions and are caused there, both by tectonic faults and the movement of magma in volcanoes. Such earthquakes can serve as an early warning of volcanic eruptions, as during the Mount St. Helens eruption of 1980. Earthquake swarms can serve as markers for the location of the flowing magma throughout the volcanoes. These swarms can be recorded by seismometers and tiltmeters (a device that measures ground slope) and used as sensors to predict imminent or upcoming eruptions.

RUPTURE DYNAMICS

A tectonic earthquake begins by an initial rupture at a point on the fault surface, a process known as nucleation. The scale of the nucleation zone is uncertain, with some evidence, such as the rupture dimensions of the smallest earthquakes, suggesting that it is smaller than 100 m while other evidence, such as a slow component revealed by low-frequency spectra of some earthquakes, suggest that it is larger. The possibility that the nucleation involves some sort of preparation process is supported by the observation that about 40% of earthquakes are preceded by foreshocks. Once the rupture has initiated, it begins to propagate along the fault surface. The mechanics of this process are poorly understood, partly because it is difficult to recreate the high sliding velocities in a laboratory. Also the effects of strong ground motion make it very difficult to record information close to a nucleation zone.

Rupture propagation is generally modeled using a fracture mechanics approach, likening the rupture to a propagating mixed mode shear crack. The rupture velocity is a function of the fracture energy in the volume around the crack tip, increasing with decreasing fracture energy. The velocity of rupture propagation is orders of magnitude faster than the displacement velocity across the fault. Earthquake ruptures typically propagate at velocities that are in the range 70–90% of the S-wave velocity, and this is independent of earthquake size. A small subset of earthquake ruptures appear to have propagated at speeds greater than the S-wave velocity. These supershear earthquakes have all been observed during large strike-slip events. The unusually wide zone of coseismic damage caused by the 2001 Kunlun earthquake has been attributed to the effects of the sonic boom developed in such earthquakes. Some earthquake ruptures travel at unusually low velocities and are referred to as slow earthquakes. A particularly dangerous form of slow earthquake is the tsunami earthquake, observed where the relatively low felt intensities, caused by the slow propagation speed of some great earthquakes, fail to alert the population of the neighbouring coast, as in the 1896 Meiji-Sanriku earthquake.

Tidal Forces

Research work has shown a robust correlation between small tidally induced forces and non-volcanic tremor activity.

Earthquake Clusters

Most earthquakes form part of a sequence, related to each other in terms of location and time. Most earthquake clusters consist of small tremors that cause little to no damage, but there is a theory that earthquakes can recur in a regular pattern.

Aftershocks

An aftershock is an earthquake that occurs after a previous earthquake, the mainshock. An aftershock is in the same region of the main shock but always of a smaller magnitude. If an aftershock is larger than the main shock, the aftershock is redesignated as the main shock and the original main shock is redesignated as a foreshock. Aftershocks are formed as the crust around the displaced fault plane adjusts to the effects of the main shock.

Earthquake Swarms

Earthquake swarms are sequences of earthquakes striking in a specific area within a short period of time. They are different from earthquakes followed by a series of aftershocks by the fact that no single earthquake in the sequence is obviously the main shock, therefore none have notable higher magnitudes than the other. An example of an earthquake swarm is the 2004 activity at Yellowstone National Park. In August 2012, a swarm of earthquakes shook Southern California's Imperial Valley, showing the most recorded activity in the area since the 1970s.

Earthquake Storms

Sometimes a series of earthquakes occur in a sort of earthquake storm, where the earthquakes strike a fault in clusters, each triggered by the shaking or stress redistribution of the previous earthquakes. Similar to aftershocks but on adjacent segments of fault, these storms occur over the course of years, and with some of the later earthquakes as damaging as the early ones. Such a pattern was observed in the sequence of about a dozen earthquakes that struck the North Anatolian Fault in Turkey in the 20th century and has been inferred for older anomalous clusters of large earthquakes in the Middle East.

Size and Frequency of Occurrence

It is estimated that around 500, 000 earthquakes occur each year, detectable with current instrumentation. About 100, 000 of these can be felt. Minor earthquakes occur nearly constantly around the world in places like California and Alaska in the U. S., as well as in El Salvador, Mexico, Guatemala, Chile, Peru, Indonesia, Iran, Pakistan, the Azores in Portugal, Turkey, New Zealand, Greece, Italy, India, Nepal and Japan, but earthquakes can occur almost anywhere, including Downstate New York, England, and Australia. Larger earthquakes occur less frequently, the relationship being exponential; for example, roughly ten times as many earthquakes larger than magnitude 4 occur in a particular time period than earthquakes larger than magnitude 5. In the (low seismicity) United Kingdom, for example, it has been calculated that the average recurrences are: an earthquake of 3. 7–4. 6 every year, an earthquake of 4. 7–5. 5 every 10 years, and an earthquake of 5. 6 or larger every 100 years. This is an example of the Gutenberg–Richter law.

The number of seismic stations has increased from about 350 in 1931 to many thousands today. As a result, many more earthquakes are reported than in the past, but this is because of the vast improvement in instrumentation, rather than an increase in the number of earthquakes. The United States Geological Survey estimates that, since 1900, there have been an average of 18 major earthquakes (magnitude 7. 0–7. 9) and one great earthquake (magnitude 8. 0 or greater) per year, and that this average has been relatively stable. In recent years, the number of major earthquakes per year has decreased, though this is probably a statistical fluctuation rather than a systematic trend. More detailed statistics on the size and frequency of earthquakes is available from the United States Geological Survey (USGS). A recent increase in the number of major earthquakes has been noted, which could be explained by a cyclical pattern of periods of intense tectonic activity, interspersed with longer periods of low-intensity. However, accurate recordings of earthquakes only began in the early 1900s, so it is too early to categorically state that this is the case.

Most of the world's earthquakes (90%, and 81% of the largest) take place in the 40, 000 km long, horseshoe-shaped zone called the circum-Pacific seismic belt, known as the Pacific Ring of Fire, which for the most part bounds the Pacific Plate. Massive earthquakes tend to occur along other plate boundaries, too, such as along the Himalayan Mountains.

With the rapid growth of mega-cities such as Mexico City, Tokyo and Tehran, in areas of high seismic risk, some seismologists are warning that a single quake may claim the lives of up to 3 million people.

Induced seismicity

While most earthquakes are caused by movement of the Earth's tectonic plates, human activity can also produce earthquakes. Four main activities contribute to this phenomenon: storing large amounts of water behind a dam (and possibly building an extremely heavy building), drilling and injecting liquid into wells, and by coal mining and oil drilling. Perhaps the best known example is the 2008 Sichuan earthquake in China's Sichuan Province in May; this tremor resulted in 69, 227 fatalities and is the 19th deadliest earthquake of all time. The Zipingpu Dam is believed to have fluctuated the pressure of the fault 1, 650 feet (503 m) away; this pressure probably increased the power of the earthquake and accelerated the rate of movement for the fault. The greatest earthquake in Australia's history is also claimed to be induced by humanity, through coal mining. The city of Newcastle was built over a large sector of coal mining areas. The earthquake has been reported to be spawned from a fault that reactivated due to the millions of tonnes of rock removed in the mining process.

Measuring and locating earthquakes

Earthquakes can be recorded by seismometers up to great distances, because seismic waves travel through the whole Earth's interior. The absolute magnitude of a quake is conventionally reported by numbers on the moment magnitude scale (formerly Richter scale, magnitude 7 causing serious damage over large areas), whereas the felt magnitude is reported using the modified Mercalli intensity scale (intensity II–XII).

Every tremor produces different types of seismic waves, which travel through rock with different velocities:

- ◉ Longitudinal P-waves (shock- or pressure waves)
- ◉ Transverse S-waves (both body waves)
- ◉ Surface waves — (Rayleigh and Love waves)

Propagation velocity of the seismic waves ranges from approx. 3 km/s up to 13 km/s, depending on the density and elasticity of the medium. In the Earth's interior the shock- or P waves travel much faster than the S waves (approx. relation 1. 7 : 1). The differences in travel time from the epicentre to the observatory are a measure of the distance and can be used to image both sources of quakes and structures within the Earth. Also the depth of the hypocenter can be computed roughly.

In solid rock P-waves travel at about 6 to 7 km per second; the velocity increases within the deep mantle to ~13 km/s. The velocity of S-waves ranges from 2–3 km/s in light sediments and 4–5 km/s in the Earth's crust up to 7 km/s in the deep mantle. As a consequence, the first waves of a distant earthquake arrive at an observatory via the Earth's mantle.

On average, the kilometer distance to the earthquake is the number of seconds between the P and S wave times 8. Slight deviations are caused by inhomogeneities of subsurface structure. By such analyses of seismograms the Earth's core was located in 1913 by Beno Gutenberg.

Earthquakes are not only categorized by their magnitude but also by the place where they occur. The world is divided into 754 Flinn–Engdahl regions (F-E regions), which are based on political and geographical boundaries as well as seismic activity. More active zones are divided into smaller F-E regions whereas less active zones belong to larger F-E regions.

Standard reporting of earthquakes includes its magnitude, date and time of occurrence, geographic coordinates of its epicenter, depth of the epicenter, geographical region, distances to population centers, location uncertainty, a number of parameters that are included in USGS earthquake reports (number of stations reporting, number of observations, etc.), and a unique event ID.

EFFECTS OF EARTHQUAKES

The effects of earthquakes include, but are not limited to, the following:

Shaking and Ground Rupture

Shaking and ground rupture are the main effects created by earthquakes, principally resulting in more or less severe damage to buildings and other rigid structures. The severity of the local effects depends on the complex combination of the earthquake magnitude, the distance from the epicenter, and the local geological and geomorphological conditions, which may amplify or reduce wave propagation. The ground-shaking is measured by ground acceleration.

Specific local geological, geomorphological, and geostructural features can induce high levels of shaking on the ground surface even from low-intensity earthquakes. This effect is called site or local amplification. It is principally due to the transfer of the seismic motion from hard deep soils to soft superficial soils and to effects of seismic energy focalization owing to typical geometrical setting of the deposits.

Ground rupture is a visible breaking and displacement of the Earth's surface along the trace of the fault, which may be of the order of several metres in the case of major earthquakes. Ground rupture is a major risk for large engineering structures such as dams, bridges and nuclear power stations and requires careful mapping of existing faults to identify any which are likely to break the ground surface within the life of the structure.

Landslides and Avalanches

Earthquakes, along with severe storms, volcanic activity, coastal wave attack, and wildfires, can produce slope instability leading to landslides, a major geological hazard. Landslide danger may persist while emergency personnel are attempting rescue.

Fires

Earthquakes can cause fires by damaging electrical power or gas lines. In the event of water mains rupturing and a loss of pressure, it may also become difficult to stop the spread of a fire once it has started. For example, more deaths in the 1906 San Francisco earthquake were caused by fire than by the earthquake itself.

Soil Liquefaction

Soil liquefaction occurs when, because of the shaking, water-saturated granular material (such as sand) temporarily loses its strength and transforms from a solid to a liquid. Soil liquefaction may cause rigid structures, like buildings and bridges, to tilt or sink into the liquefied deposits. For example, in the 1964 Alaska earthquake, soil liquefaction caused many buildings to sink into the ground, eventually collapsing upon themselves.

Tsunami

Tsunamis are long-wavelength, long-period sea waves produced by the sudden or abrupt movement of large volumes of water. In the open ocean the distance between wave crests can surpass 100 kilometers (62 mi), and the wave periods can vary from five minutes to one hour. Such tsunamis travel 600-800 kilometers per hour (373–497 miles per hour), depending on water depth. Large waves produced by an earthquake or a submarine landslide can overrun nearby coastal areas in a matter of minutes. Tsunamis can also travel thousands of kilometers across open ocean and wreak destruction on far shores hours after the earthquake that generated them.

Ordinarily, subduction earthquakes under magnitude 7. 5 on the Richter scale do not cause tsunamis, although some instances of this have been recorded. Most destructive tsunamis are caused by earthquakes of magnitude 7. 5 or more.

Floods

A flood is an overflow of any amount of water that reaches land. Floods occur usually when the volume of water within a body of water, such as a river or lake, exceeds the total capacity of the formation, and as a result some of the water flows or sits outside of the normal perimeter of the body. However, floods may be secondary effects of earthquakes, if dams are damaged. Earthquakes may cause landslips to dam rivers, which collapse and cause floods.

The terrain below the Sarez Lake in Tajikistan is in danger of catastrophic flood if the landslide dam formed by the earthquake, known as the Usoi Dam, were to fail during a future earthquake. Impact projections suggest the flood could affect roughly 5 million people.

STRUCTURAL GEOLOGY 12

Structural geology is the study of the three-dimensional distribution of rock units with respect to their deformational histories. The primary goal of structural geology is to use measurements of present-day rock geometries to uncover information about the history of deformation (strain) in the rocks, and ultimately, to understand the stress field that resulted in the observed strain and geometries. This understanding of the dynamics of the stress field can be linked to important events in the regional geologic past; a common goal is to understand the structural evolution of a particular area with respect to regionally widespread patterns of rock deformation (e. g., mountain building, rifting) due to plate tectonics.

USE AND IMPORTANCE

The study of geologic structures has been of prime importance in economic geology, both petroleum geology and mining geology. Folded and faulted rock strata commonly form traps for the accumulation and concentration of fluids such as petroleum and natural gas. Faulted and structurally complex areas are notable as permeable zones for hydrothermal fluids and the resulting concentration areas for base and precious metal ore deposits. Veins of minerals containing various metals commonly occupy faults and fractures in structurally complex areas. These structurally fractured and faulted zones often occur in association with intrusive igneous rocks. They often also occur around geologic reefcomplexes and collapse features such as ancient sinkholes. Deposits of gold, silver, copper, lead, zinc, and other metals, are commonly located in structurally complex areas.

Structural geology is a critical part of engineering geology, which is concerned with the physical and mechanical properties of natural rocks. Structural fabrics and defects such as faults, folds, foliations and joints are internal weaknesses of rocks which may affect the stability of human engineered structures such as dams, road cuts, open pit mines and underground mines or road tunnels.

Geotechnical risk, including earthquake risk can only be investigated by inspecting a combination of structural geology and geomorphology. In addition areas of karst landscapes which are underlain by underground caverns and potential sinkholes or collapse features are of importance for these scientists. In addition, areas of steep slopes are potential collapse or landslide hazards.

Environmental geologists and hydrogeologists or hydrologists need to understand structural geology because structures are sites of groundwater flow and penetration, which may affect, for instance, seepage of toxic substances from waste dumps, or seepage of salty water into aquifers.

Plate tectonics is a theory developed during the 1960s which describes the movement of continents by way of the separation and collision of crustal plates. It is in a sense structural geology on a planet scale, and is used throughout structural geology as a framework to analyze and understand global, regional, and local scale features.

Methods

Structural geologists use a variety of methods to (first) measure rock geometries, (second) reconstruct their deformational histories, and (third) calculate the stress field that resulted in that deformation.

Geometries

Primary data sets for structural geology are collected in the field. Structural geologists measure a variety of planar features (bedding planes, foliation planes, fold axial planes, fault planes, and joints), and linear features (stretching lineations, in which minerals are ductily extended; fold axes; and intersection lineations, the trace of a planar feature on another planar surface).

Measurement Conventions

The inclination of a planar structure in geology is measured by strike and dip. The strike is the line of intersection between the planar feature and a horizontal plane, taken according to the right hand convention, and the dip is the magnitude of the inclination, below horizontal, at right angles to strike. For example; striking 25 degrees East of North, dipping 45 degrees Southeast, recorded as N25E, 45SE.

Alternatively, dip and dip direction may be used as this is absolute. Dip direction is measured in 360 degrees, generally clockwise from North. For example, a dip of 45 degrees towards 115 degrees azimuth, recorded as 45/115. Note that this is the same as above.

The term hade is occasionally used and is the deviation of a plane from vertical i. e. (90°-dip).

Fold axis plunge is measured in dip and dip direction (strictly, plunge and azimuth of plunge). The orientation of a fold axial plane is measured in strike and dip or dip and dip direction.

Lineations are measured in terms of dip and dip direction, if possible. Often lineations occur expressed on a planar surface and can be difficult to measure directly. In this case, the lineation may be measured from the horizontal as a rakeor pitch upon the surface.

Rake is measured by placing a protractor flat on the planar surface, with the flat edge horizontal and measuring the angle of the lineation clockwise from horizontal. The orientation of the lineation can then be calculated from the rake and strike-dip information of the plane it was measured from, using a stereographic projection.

If a fault has lineations formed by movement on the plane, e. g. ; slickensides, this is recorded as a lineation, with a rake, and annotated as to the indication of throw on the fault.

Generally it is easier to record strike and dip information of planar structures in dip/dip direction format as this will match all the other structural information you may be recording about folds, lineations, etc., although there is an advantage to using different formats that discriminate between planar and linear data.

Plane, fabric, fold and deformation conventions

The convention for analysing structural geology is to identify the planar structures, often called planar fabrics because this implies a texturalformation, the linear structures and, from analysis of these, unravel deformations.

Planar structures are named according to their order of formation, with original sedimentary layering the lowest at S0. Often it is impossible to identify S0 in highly deformed rocks, so numbering may be started at an arbitrary number or given a letter (SA, for instance). In cases where there is a bedding-plane foliation caused by burial metamorphism or diagenesis this may be enumerated as S0a.

If there are folds, these are numbered as F1, F2, etc. Generally the axial plane foliation or cleavage of a fold is created during folding, and the number convention should match. For example, an F2 fold should have an S2 axial foliation.

Deformations are numbered according to their order of formation with the letter D denoting a deformation event. For example D1, D2, D3. Folds and foliations, because they are formed by deformation events, should correlate with these events. For example an F2 fold, with an S2axial plane foliation would be the result of a D2 deformation.

Metamorphic events may span multiple deformations. Sometimes it is useful to identify them similarly to the structural features for which they are responsible, e. g. ; M2. This may be possible by observing porphyroblast formation in cleavages of known deformation age, by identifying metamorphic mineral assemblages created by different events, or via geochronology.

Intersection lineations in rocks, as they are the product of the intersection of two planar structures, are named according to the two planar structures from which they are formed. For instance, the intersection lineation of a S1 cleavage and bedding is the L1-0 intersection lineation (also known as the cleavage-bedding lineation).

Stretching lineations may be difficult to quantify, especially in highly stretched ductile rocks where minimal foliation information is preserved. Where possible, when correlated with deformations (as few are formed in folds, and many are not strictly associated with planar foliations), they may be identified similar to planar surfaces and folds, e. g. ; L1, L2. For convenience some geologists prefer to annotate them with a subscript S, for example Ls1 to differentiate them from intersection lineations, though this is generally redundant.

STEREOGRAPHIC PROJECTIONS

Stereographic projection of structural strike and dip measurements is a powerful method for analyzing the nature and orientation of deformation stresses, lithological units and penetrative fabrics.

Rock Macro-structures

On a large scale, structural geology is the study of the three dimensional relationships of stratigraphic units to one another within terranes of rock or within geological regions. This branch of structural geology deals mainly with the orientation, deformation and relationships of stratigraphy (bedding), which may have been faulted, folded or given a foliation by some tectonic event. This is mainly a geometric science, from which cross sections and three dimensional block models of rocks, regions, terranes and parts of the Earth's crust can be generated.

Study of regional structure is important in understanding orogeny, plate tectonics and more specifically in the oil, gas and mineral exploration industries as structures such as faults, folds and unconformities are primary controls on ore mineralisation and oil traps.

Modern regional structure is being investigated using seismic tomography and seismic reflection in three dimensions, providing unrivaled images of the Earth's interior, its faults and the deep crust. Further information from geophysics such as gravity and airborne magnetics can provide information on the nature of rocks imaged in the deep crust.

Rock microstructures

Rock microstructure or texture of rocks is studied by structural geologists on a small scale to provide detailed information mainly aboutmetamorphic rocks and some features of sedimentary rocks, most often if they have been folded.

Textural study involves measurement and characterisation of foliations, crenulations, metamorphic minerals, and timing relationships between these structural features and mineralogical features.

Usually this involves collection of hand specimens, which may be cut to provide petrographic thin sections which are analysed under apetrographic microscope.

Microstructural analysis finds application also in multi-scale statistical analysis, aimed to analyze some rock features showing scale invariance.

Kinematics

Geologists use their measurements of rock geometries to understand histories of strain in the rocks. Strain can take the form of brittle faulting and ductile folding and shearing. Brittle deformation takes place in the shallow crust, and ductile deformation takes place in the deeper crust, where temperatures and pressures are higher.

Stress Fields

By understanding the constitutive relationships between stress and strain in rocks, geologists can translate the observed patterns of rock deformation into a stress field during the geologic past. The following list of features are typically used to determine stress fields from deformational structures.

- In perfectly brittle rocks, faulting occurs at 30° to the greatest compressional stress. (Byerlee's Law)

- The greatest compressive stress is normal to fold axial planes.

PLATE TECTONICS

13

Plate tectonics is a scientific theory that describes the large-scale motion of Earth'slithosphere. This theoretical model builds on the concept of continental drift which was developed during the first few decades of the 20th century. The geoscientific community accepted the theory after the concepts of seafloor spreading were later developed in the late 1950s and early 1960s.

The lithosphere, which is the rigid outermost shell of a planet (on Earth, the crust and upper mantle), is broken up into tectonic plates. On Earth, there are seven or eight major plates (depending on how they are defined) and many minor plates. Where plates meet, their relative motion determines the type of boundary; convergent, divergent, ortransform. Earthquakes, volcanic activity, mountain-building, and oceanic trench formation occur along these plate boundaries. The lateral relative movement of the plates typically varies from zero to 100 mm annually.

Tectonic plates are composed of oceanic lithosphere and thicker continental lithosphere, each topped by its own kind of crust. Along convergent boundaries, subduction carries plates into the mantle; the material lost is roughly balanced by the formation of new (oceanic) crust along divergent margins by seafloor spreading. In this way, the total surface of the globe remains the same. This prediction of plate tectonics is also referred to as the conveyor belt principle. Earlier theories (that still have some supporters) propose gradual shrinking (contraction) or gradual expansion of the globe.

Tectonic plates are able to move because the Earth's lithosphere has greater strength than the underlying asthenosphere. Lateral density variations in the mantle result inconvection. Plate movement is thought to be driven by a combination of the motion of the seafloor away from the spreading ridge (due to variations in topography and density of the crust, which result in differences in gravitational forces) and drag, with downwardsuction, at the subduction zones. Another explanation lies in the different forces generated by the rotation of the globe and the tidal forces of the Sun and Moon. The relative importance of each of these factors and their relationship to each other is unclear, and still the subject of much debate.

KEY PRINCIPLES

The outer layers of the Earth are divided into the lithosphere and asthenosphere. This is based on differences in mechanical properties and in the method for the transfer of heat. Mechanically, the lithosphere is cooler and more rigid, while the asthenosphere is hotter and flows more easily. In terms of heat transfer, the lithosphere loses heat by conduction, whereas the asthenosphere also transfers heat by convection and has a nearly adiabatic temperature gradient. This division should not be confused with the chemical subdivision of these same layers into the mantle (comprising both the asthenosphere and the mantle portion of the lithosphere) and the crust: a given piece of mantle may be part of the lithosphere or the asthenosphere at different times depending on its temperature and pressure.

The key principle of plate tectonics is that the lithosphere exists as separate and distinct tectonic plates, which ride on the fluid-like (visco-elastic solid) asthenosphere. Plate motions range up to a typical 10–40 mm/year (Mid-Atlantic Ridge; about as fast as fingernails grow), to about 160 mm/year (Nazca Plate; about as fast as hair grows). The driving mechanism behind this movement is described below.

Tectonic lithosphere plates consist of lithospheric mantle overlain by either or both of two types of crustal material: oceanic crust (in older texts called sima from silicon and magnesium) and continental crust (sial from silicon and aluminium). Average oceanic lithosphere is typically 100 km (62 mi) thick; its thickness is a function of its age: as time passes, it conductively cools and subjacent cooling mantle is added to its base. Because it is formed at mid-ocean ridges and spreads outwards, its thickness is therefore a function of its distance from the mid-ocean ridge where it was formed. For a typical distance that oceanic lithosphere must travel before being subducted, the thickness varies from about 6 km (4 mi) thick at mid-ocean ridges to greater than 100 km (62 mi) at subduction zones; for shorter or longer distances, the subduction zone (and therefore also the mean) thickness becomes smaller or larger, respectively. Continental lithosphere is typically ~200 km thick, though this varies considerably between basins, mountain ranges, and stable cratonic interiors of continents. The two types of crust also differ in thickness, with continental crust being considerably thicker than oceanic (35 km vs. 6 km).

The location where two plates meet is called a plate boundary. Plate boundaries are commonly associated with geological events such as earthquakes and the creation of topographic features such as mountains, volcanoes, mid-ocean ridges, and oceanic trenches. The majority of the world's active volcanoes occur along plate boundaries, with the Pacific Plate's Ring of Fire being the most active and widely known today.

These boundaries are discussed in further detail below. Some volcanoes occur in the interiors of plates, and these have been variously attributed to internal plate deformation and to mantle plumes.

As explained above, tectonic plates may include continental crust or oceanic crust, and most plates contain both. For example, the African Plate includes the continent and parts of the floor of the Atlantic and Indian Oceans. The distinction between oceanic crust and continental crust is based on their modes of formation. Oceanic crust is formed at sea-floor spreading centers, and continental crust is formed through arc volcanism and accretion of terranes through tectonic processes, though some of these terranes may contain ophiolite sequences, which are pieces of oceanic crust considered to be part of the continent when they exit the standard cycle of formation and spreading centers and subduction beneath continents. Oceanic crust is also denser than continental crust owing to their different compositions. Oceanic crust is denser because it has less silicon and more heavier elements ("mafic") than continental crust ("felsic"). As a result of this density stratification, oceanic crust generally lies below sea level (for example most of the Pacific Plate), while continental crust buoyantly projects above sea level.

Types of plate boundaries

Three types of plate boundaries exist, with a fourth, mixed type, characterized by the way the plates move relative to each other. They are associated with different types of surface phenomena. The different types of plate boundaries are:

- ◉ Transform boundaries (Conservative) occur where two lithospheric plates slide, or perhaps more accurately, grind past each other alongtransform faults, where plates are neither created nor destroyed. The relative motion of the two plates is either sinistral (left side toward the observer) or dextral (right side toward the observer). Transform faults occur across a spreading center. Strong earthquakes can occur along a fault. The San Andreas Fault in California is an example of a transform boundary exhibiting dextral motion.

- ◉ Divergent boundaries (Constructive) occur where two plates slide apart from each other. At zones of ocean-to-ocean rifting, divergent boundaries form by seafloor spreading, allowing for the formation of new ocean basin. As the continent splits, the ridge forms at the spreading center, the ocean basin expands, and finally, the plate area increases causing many small volcanoes and/or shallow earthquakes. At zones of continent-to-continent rifting, divergent boundaries may cause new ocean basin to form as the continent splits, spreads, the central rift collapses, and ocean fills the basin. Active zones of Mid-ocean ridges (e. g., Mid-Atlantic Ridge and East Pacific Rise), and continent-to-continent rifting (such as Africa's East African Rift and Valley, Red Sea) are examples of divergent boundaries.

- Convergent boundaries (Destructive) (or active margins) occur where two plates slide toward each other to form either a subductionzone (one plate moving underneath the other) or a continental collision. At zones of ocean-to-continent subduction (e. g., Western South America, and Cascade Mountains in Western United States), the dense oceanic lithosphere plunges beneath the less dense continent. Earthquakes then trace the path of the downward-moving plate as it descends into asthenosphere, a trench forms, and as the subducted plate partially melts, magma rises to form continental volcanoes. At zones of ocean-to-ocean subduction (e. g., theAndes mountain range in South America, Aleutian islands, Mariana islands, and the Japanese island arc), older, cooler, denser crust slips beneath less dense crust. This causes earthquakes and a deep trench to form in an arc shape. The upper mantle of the subducted plate then heats and magma rises to form curving chains of volcanic islands. Deep marine trenches are typically associated with subduction zones, and the basins that develop along the active boundary are often called "foreland basins". The subducting slabcontains many hydrous minerals which release their water on heating. This water then causes the mantle to melt, producing volcanism. Closure of ocean basins can occur at continent-to-continent boundaries (e. g., Himalayas and Alps): collision between masses of granitic continental lithosphere; neither mass is subducted; plate edges are compressed, folded, uplifted.

- Plate boundary zones occur where the effects of the interactions are unclear, and the boundaries, usually occurring along a broad belt, are not well defined and may show various types of movements in different episodes.

DRIVING FORCES OF PLATE MOTION

Plate tectonics is basically a kinematic phenomenon. Scientists agree on the observation and deduction that the plates have moved with respect to one another but continue to debate as to how and when. A major question remains as to what geodynamic mechanism motors plate movement. Here, science diverges in different theories.

It is generally accepted that tectonic plates are able to move because of the relative density of oceanic lithosphere and the relative weakness of the asthenosphere. Dissipation of heat from the mantle is acknowledged to be the original source of the energy required to drive plate tectonics through convection or large scale upwelling and doming. The current view, though still a matter of some debate, asserts that as a consequence, a powerful source of plate motion is generated due to the excess density of the oceanic lithosphere sinking in subduction zones. When the new crust forms at mid-ocean ridges, this oceanic lithosphere is initially less dense than the underlying asthenosphere, but it becomes denser with age as it conductively cools and thickens.

The greater density of old lithosphere relative to the underlying asthenosphere allows it to sink into the deep mantle at subduction zones, providing most of the driving force for plate movement. The weakness of the asthenosphere allows the tectonic plates to move easily towards a subduction zone. Although subduction is believed to be the strongest force driving plate motions, it cannot be the only force since there are plates such as the North American Plate which are moving, yet are nowhere being subducted. The same is true for the enormous Eurasian Plate. The sources of plate motion are a matter of intensive research and discussion among scientists. One of the main points is that the kinematic pattern of the movement itself should be separated clearly from the possible geodynamic mechanism that is invoked as the driving force of the observed movement, as some patterns may be explained by more than one mechanism. In short, the driving forces advocated at the moment can be divided into three categories based on the relationship to the movement: mantle dynamics related, gravity related (mostly secondary forces), and Earth rotation related.

Driving forces related to mantle dynamics

For much of the last quarter century, the leading theory of the driving force behind tectonic plate motions envisaged large scale convection currents in the upper mantle which are transmitted through the asthenosphere. This theory was launched by Arthur Holmes and some forerunners in the 1930s and was immediately recognized as the solution for the acceptance of the theory as originally discussed in the papers of Alfred Wegener in the early years of the century. However, despite its acceptance, it was long debated in the scientific community because the leading ("fixist") theory still envisaged a static Earth without moving continents up until the major breakthroughs of the early sixties.

Two- and three-dimensional imaging of Earth's interior (seismic tomography) shows a varying lateral density distribution throughout the mantle. Such density variations can be material (from rock chemistry), mineral (from variations in mineral structures), or thermal (through thermal expansion and contraction from heat energy). The manifestation of this varying lateral density is mantle convection from buoyancy forces.

How mantle convection directly and indirectly relates to plate motion is a matter of ongoing study and discussion in geodynamics. Somehow, this energy must be transferred to the lithosphere for tectonic plates to move. There are essentially two types of forces that are thought to influence plate motion: friction and gravity.

- ◉ Basal drag (friction): Plate motion driven by friction between the convection currents in the asthenosphere and the more rigid overlying lithosphere.

- ◉ Slab suction (gravity): Plate motion driven by local convection currents that exert a downward pull on plates in subduction zones at ocean trenches. Slab suction may occur in a geodynamic setting where basal tractions continue to act on the plate as it dives into the mantle (although perhaps to a greater extent acting on both the under and upper side of the slab).

Lately, the convection theory has been much debated as modern techniques based on 3D seismic tomography still fail to recognize these predicted large scale convection cells. Therefore, alternative views have been proposed:

In the theory of plume tectonics developed during the 1990s, a modified concept of mantle convection currents is used. It asserts that super plumes rise from the deeper mantle and are the drivers or substitutes of the major convection cells. These ideas, which find their roots in the early 1930s with the so-called "fixistic" ideas of the European and Russian Earth Science Schools, find resonance in the modern theories which envisage hot spots/mantle plumes which remain fixed and are overridden by oceanic and continental lithosphere plates over time and leave their traces in the geological record (though these phenomena are not invoked as real driving mechanisms, but rather as modulators). Modern theories that continue building on the older mantle doming concepts and see plate movements as a secondary phenomena are beyond the scope of this page and are discussed elsewhere (for example on the plume tectonics page).

Another theory is that the mantle flows neither in cells nor large plumes but rather as a series of channels just below the Earth's crust, which then provide basal friction to the lithosphere. This theory, called "surge tectonics", became quite popular in geophysics and geodynamics during the 1980s and 1990s.

Driving forces related to gravity

Forces related to gravity are usually invoked as secondary phenomena within the framework of a more general driving mechanism such as the various forms of mantle dynamics described above. Gravitational sliding away from a spreading ridge: According to many authors, plate motion is driven by the higher elevation of plates at ocean ridges. As oceanic lithosphere is formed at spreading ridges from hot mantle material, it gradually cools and thickens with age (and thus adds distance from the ridge). Cool oceanic lithosphere is significantly denser than the hot mantle material from which it is derived and so with increasing thickness it gradually subsides into the mantle to compensate the greater load. The result is a slight lateral incline with increased distance from the ridge axis.

This force is regarded as a secondary force and is often referred to as "ridge push". This is a misnomer as nothing is "pushing" horizontally and tensional features are dominant along ridges. It is more accurate to refer to this mechanism as gravitational sliding as variable topography across the totality of the plate can vary considerably and the topography of spreading ridges is only the most prominent feature. Other mechanisms generating this gravitational secondary force include flexural bulging of the lithosphere before it dives underneath an adjacent plate which produces a clear topographical feature that can offset, or at least affect, the influence of topographical ocean ridges, and mantle plumes and hot spots, which are postulated to impinge on the underside of tectonic plates. Slab-pull: Current scientific opinion is that the asthenosphere is insufficiently competent or rigid to directly cause motion by friction along the base of the lithosphere. Slab pull is therefore most widely thought to be the greatest force acting on the plates. In this current understanding, plate motion is mostly driven by the weight of cold, dense plates sinking into the mantle at trenches. Recent models indicate that trench suction plays an important role as well. However, as the North American Plate is nowhere being subducted, yet it is in motion presents a problem. The same holds for the African, Eurasian, and Antarctic plates.

Gravitational sliding away from mantle doming: According to older theories, one of the driving mechanisms of the plates is the existence of large scale asthenosphere/mantle domes which cause the gravitational sliding of lithosphere plates away from them. This gravitational sliding represents a secondary phenomenon of this basically vertically oriented mechanism. This can act on various scales, from the small scale of one island arc up to the larger scale of an entire ocean basin.

Driving forces related to Earth rotation

Alfred Wegener, being a meteorologist, had proposed tidal forces and pole flight force as the main driving mechanisms behind continental drift; however, these forces were considered far too small to cause continental motion as the concept then was of continents plowing through oceanic crust. Therefore, Wegener later changed his position and asserted that convection currents are the main driving force of plate tectonics in the last edition of his book in 1929.

However, in the plate tectonics context (accepted since the seafloor spreading proposals of Heezen, Hess, Dietz, Morley, Vine, and Matthews during the early 1960s), oceanic crust is suggested to be in motion with the continents which caused the proposals related to Earth rotation to be reconsidered. In more recent literature, these driving forces are:

⊙ Tidal drag due to the gravitational force the Moon (and the Sun) exerts on the crust of the Earth

⊙ Shear strain of the Earth globe due to N-S compression related to its rotation and modulations;

⊙ Pole flight force: equatorial drift due to rotation and centrifugal effects: tendency of the plates to move from the poles to the equator ("Polflucht");

⊙ The Coriolis effect acting on plates when they move around the globe;

- ◉ Global deformation of the geoid due to small displacements of rotational pole with respect to the Earth's crust;

- ◉ Other smaller deformation effects of the crust due to wobbles and spin movements of the Earth rotation on a smaller time scale.

For these mechanisms to be overall valid, systematic relationships should exist all over the globe between the orientation and kinematics of deformation and the geographical latitudinal and longitudinal grid of the Earth itself. Ironically, these systematic relations studies in the second half of the nineteenth century and the first half of the twentieth century underline exactly the opposite: that the plates had not moved in time, that the deformation grid was fixed with respect to the Earth equator and axis, and that gravitational driving forces were generally acting vertically and caused only local horizontal movements (the so-called pre-plate tectonic, "fixist theories"). Later studies (discussed below on this page), therefore, invoked many of the relationships recognized during this pre-plate tectonics period to support their theories.

Of the many forces discussed in this paragraph, tidal force is still highly debated and defended as a possible principle driving force of plate tectonics. The other forces are only used in global geodynamic models not using plate tectonics concepts (therefore beyond the discussions treated in this section) or proposed as minor modulations within the overall plate tectonics model.

In 1973, George W. Moore of the USGS and R. C. Bostrom presented evidence for a general westward drift of the Earth's lithosphere with respect to the mantle. He concluded that tidal forces (the tidal lag or "friction") caused by the Earth's rotation and the forces acting upon it by the Moon are a driving force for plate tectonics. As the Earth spins eastward beneath the moon, the moon's gravity ever so slightly pulls the Earth's surface layer back westward, just as proposed by Alfred Wegener. In a more recent 2006 study, scientists reviewed and advocated these earlier proposed ideas. It has also been suggested recently in Lovett (2006) that this observation may also explain whyVenus and Mars have no plate tectonics, as Venus has no moon and Mars' moons are too small to have significant tidal effects on the planet. In a recent paper, it was suggested that, on the other hand, it can easily be observed that many plates are moving north and eastward, and that the dominantly westward motion of the Pacific ocean basins derives simply from the eastward bias of the Pacific spreading center (which is not a predicted manifestation of such lunar forces). In the same paper the authors admit, however, that relative to the lower mantle, there is a slight westward component in the motions of all the plates. They demonstrated though that the westward drift, seen only for the past 30 Ma, is attributed to the increased dominance of the steadily growing and accelerating Pacific plate. The debate is still open.

Relative significance of each driving force mechanism

The actual vector of a plate's motion is a function of all the forces acting on the plate; however, therein lies the problem regarding the degree to which each process contributes to the overall motion of each tectonic plate.

The diversity of geodynamic settings and the properties of each plate result from the impact of the various processes actively driving each individual plate. One method of dealing with this problem is to consider the relative rate at which each plate is moving as well as the evidence related to the significance of each process to the overall driving force on the plate.

One of the most significant correlations discovered to date is that lithospheric plates attached to downgoing (subducting) plates move much faster than plates not attached to subducting plates. The Pacific plate, for instance, is essentially surrounded by zones of subduction (the so-called Ring of Fire) and moves much faster than the plates of the Atlantic basin, which are attached (perhaps one could say 'welded') to adjacent continents instead of subducting plates. It is thus thought that forces associated with the downgoing plate (slab pull and slab suction) are the driving forces which determine the motion of plates, except for those plates which are not being subducted. The driving forces of plate motion continue to be active subjects of on-going research within geophysics and tectonophysics.

DEVELOPMENT OF THE THEORY

Summary

In line with other previous and contemporaneous proposals, in 1912 the meteorologist Alfred Wegener amply described what he called continental drift, expanded in his 1915 book The Origin of Continents and Oceansand the scientific debate started that would end up fifty years later in the theory of plate tectonics. Starting from the idea (also expressed by his forerunners) that the present continents once formed a single land mass (which was called Pangea later on) that drifted apart, thus releasing the continents from the Earth's mantle and likening them to "icebergs" of low density granite floating on a sea of denser basalt. Supporting evidence for the idea came from the dove-tailing outlines of South America's east coast and Africa's west coast, and from the matching of the rock formations along these edges. Confirmation of their previous contiguous nature also came from the fossil plants Glossopteris and Gangamopteris, and the therapsid or mammal-like reptile Lystrosaurus, all widely distributed over South America, Africa, Antarctica, India and Australia. The evidence for such an erstwhile joining of these continents was patent to field geologists working in the southern hemisphere. The South African Alex du Toit put together a mass of such information in his 1937 publication Our Wandering Continents, and went further than Wegener in recognising the strong links between the Gondwana fragments.

But without detailed evidence and a force sufficient to drive the movement, the theory was not generally accepted: the Earth might have a solid crust and mantle and a liquid core, but there seemed to be no way that portions of the crust could move around. Distinguished scientists, such as Harold Jeffreys and Charles Schuchert, were outspoken critics of continental drift.

Despite much opposition, the view of continental drift gained support and a lively debate started between "drifters" or "mobilists" (proponents of the theory) and "fixists" (opponents). During the 1920s, 1930s and 1940s, the former reached important milestones proposing thatconvection currents might have driven the plate movements, and that spreading may have occurred below the sea within the oceanic crust. Concepts close to the elements now incorporated in plate tectonics were proposed by geophysicists and geologists (both fixists and mobilists) like Vening-Meinesz, Holmes, and Umbgrove.

One of the first pieces of geophysical evidence that was used to support the movement of lithospheric plates came from paleomagnetism. This is based on the fact that rocks of different ages show a variable magnetic field direction, evidenced by studies since the mid–nineteenth century. The magnetic north and south poles reverse through time, and, especially important in paleotectonic studies, the relative position of the magnetic north pole varies through time. Initially, during the first half of the twentieth century, the latter phenomenon was explained by introducing what was called "polar wander", i. e., it was assumed that the north pole location had been shifting through time. An alternative explanation, though, was that the continents had moved (shifted and rotated) relative to the north pole, and each continent, in fact, shows its own "polar wander path". During the late 1950s it was successfully shown on two occasions that these data could show the validity of continental drift: by Keith Runcorn in a paper in 1956, and by Warren Carey in a symposium held in March 1956.

The second piece of evidence in support of continental drift came during the late 1950s and early 60s from data on the bathymetry of the deep ocean floors and the nature of the oceanic crust such as magnetic properties and, more generally, with the development of marine geology which gave evidence for the association of seafloor spreading along the mid-oceanic ridges and magnetic field reversals, published between 1959 and 1963 by Heezen, Dietz, Hess, Mason, Vine & Matthews, and Morley.

Simultaneous advances in early seismic imaging techniques in and around Wadati-Benioff zones along the trenches bounding many continental margins, together with many other geophysical (e. g. gravimetric) and geological observations, showed how the oceanic crust could disappear into the mantle, providing the mechanism to balance the extension of the ocean basins with shortening along its margins.

All this evidence, both from the ocean floor and from the continental margins, made it clear around 1965 that continental drift was feasible and the theory of plate tectonics, which was defined in a series of papers between 1965 and 1967, was born, with all its extraordinary explanatory and predictive power. The theory revolutionized the Earth sciences, explaining a diverse range of geological phenomena and their implications in other studies such as paleogeography and paleobiology.

Continental Drift

In the late 19th and early 20th centuries, geologists assumed that the Earth's major features were fixed, and that most geologic features such as basin development and mountain ranges could be explained by vertical crustal movement, described in what is called the geosynclinal theory. Generally, this was placed in the context of a contracting planet Earth due to heat loss in the course of a relatively short geological time.

It was observed as early as 1596 that the opposite coasts of the Atlantic Ocean— or, more precisely, the edges of the continental shelves—have similar shapes and seem to have once fitted together.

Since that time many theories were proposed to explain this apparent complementarity, but the assumption of a solid Earth made these various proposals difficult to accept.

The discovery of radioactivity and its associated heating properties in 1895 prompted a re-examination of the apparent age of the Earth. This had previously been estimated by its cooling rate and assumption the Earth's surface radiated like a black body. Those calculations had implied that, even if it started at red heat, the Earth would have dropped to its present temperature in a few tens of millions of years. Armed with the knowledge of a new heat source, scientists realized that the Earth would be much older, and that its core was still sufficiently hot to be liquid.

By 1915, after having published a first article in 1912, Alfred Wegener was making serious arguments for the idea of continental drift in the first edition of The Origin of Continents and Oceans. In that book (re-issued in four successive editions up to the final one in 1936), he noted how the east coast of South America and the west coast of Africa looked as if they were once attached. Wegener was not the first to note this (Abraham Ortelius, Antonio Snider-Pellegrini, Eduard Suess, Roberto Mantovani and Frank Bursley Taylor preceded him just to mention a few), but he was the first to marshal significant fossil and paleo-topographical and climatological evidence to support this simple observation (and was supported in this by researchers such as Alex du Toit). Furthermore, when the rock strata of the margins of separate continents are very similar it suggests that these rocks were formed in the same way, implying that they were joined initially. For instance, parts of Scotland and Ireland contain rocks very similar to those found in Newfoundland and New Brunswick. Furthermore, the Caledonian Mountains of Europe and parts of the Appalachian Mountains of North America are very similar in structure and lithology.

However, his ideas were not taken seriously by many geologists, who pointed out that there was no apparent mechanism for continental drift. Specifically, they did not see how continental rock could plow through the much denser rock that makes up oceanic crust. Wegener could not explain the force that drove continental drift, and his vindication did not come until after his death in 1930.

Floating Continents, Paleomagnetism, and Seismicity Zones

As it was observed early that although granite existed on continents, seafloor seemed to be composed of denser basalt, the prevailing concept during the first half of the twentieth century was that there were two types of crust, named "sial" (continental type crust) and "sima" (oceanic type crust). Furthermore, it was supposed that a static shell of strata was present under the continents. It therefore looked apparent that a layer of basalt (sial) underlies the continental rocks.

However, based on abnormalities in plumb line deflection by the Andes in Peru, Pierre Bouguer had deduced that less-dense mountains must have a downward projection into the denser layer underneath. The concept that mountains had "roots" was confirmed byGeorge B. Airy a hundred years later, during study of Himalayan gravitation, and seismic studies detected corresponding density variations. Therefore, by the mid-1950s, the question remained unresolved as to whether mountain roots were clenched in surrounding basalt or were floating on it like an iceberg.

During the 20th century, improvements in and greater use of seismic instruments such as seismographs enabled scientists to learn that earthquakes tend to be concentrated in specific areas, most notably along the oceanic trenches and spreading ridges. By the late 1920s, seismologists were beginning to identify several prominent earthquake zones parallel to the trenches that typically were inclined 40–60° from the horizontal and extended several hundred kilometers into the Earth. These zones later became known as Wadati-Benioff zones, or simply Benioff zones, in honor of the seismologists who first recognized them, Kiyoo Wadati of Japan and Hugo Benioff of the United States. The study of global seismicity greatly advanced in the 1960s with the establishment of the Worldwide Standardized Seismograph Network (WWSSN) to monitor the compliance of the 1963 treaty banning above-ground testing of nuclear weapons. The much improved data from the WWSSN instruments allowed seismologists to map precisely the zones of earthquake concentration world wide.

Meanwhile, debates developed around the phenomena of polar wander. Since the early debates of continental drift, scientists had discussed and used evidence that polar drift had occurred because continents seemed to have moved through different climatic zones during the past. Furthermore, paleomagnetic data had shown that the magnetic pole had also shifted during time. Reasoning in an opposite way, the continents might have shifted and rotated, while the pole remained relatively fixed. The first time the evidence of magnetic polar wander was used to support the movements of continents was in a paper by Keith Runcorn in 1956, and successive papers by him and his studentsTed Irving (who was actually the first to be convinced of the fact that paleomagnetism supported continental drift) and Ken Creer.

This was immediately followed by a symposium in Tasmania in March 1956. In this symposium, the evidence was used in the theory of anexpansion of the global crust. In this hypothesis the shifting of the continents can be simply explained by a large increase in size of the Earth since its formation. However, this was unsatisfactory because its supporters could offer no convincing mechanism to produce a significant expansion of the Earth. Certainly there is no evidence that the moon has expanded in the past 3 billion years; other work would soon show that the evidence was equally in support of continental drift on a globe with a stable radius.

During the thirties up to the late fifties, works by Vening-Meinesz, Holmes, Umbgrove, and numerous others outlined concepts that were close or nearly identical to modern plate tectonics theory. In particular, the English geologist Arthur Holmes proposed in 1920 that plate junctions might lie beneath the sea, and in 1928 that convection currents within the mantle might be the driving force. Often, these contributions are forgotten because:

- At the time, continental drift was not accepted.

- Some of these ideas were discussed in the context of abandoned fixistic ideas of a deforming globe without continental drift or an expanding Earth.

- They were published during an episode of extreme political and economic instability that hampered scientific communication.

- Many were published by European scientists and at first not mentioned or given little credit in the papers on sea floor spreading published by the American researchers in the 1960s.

MID-OCEANIC RIDGE SPREADING AND CONVECTION

In 1947, a team of scientists led by Maurice Ewing utilizing the Woods Hole Oceanographic Institution's research vessel Atlantis and an array of instruments, confirmed the existence of a rise in the central Atlantic Ocean, and found that the floor of the seabed beneath the layer of sediments consisted of basalt, not the granite which is the main constituent of continents. They also found that the oceanic crust was much thinner than continental crust. All these new findings raised important and intriguing questions.

The new data that had been collected on the ocean basins also showed particular characteristics regarding the bathymetry. One of the major outcomes of these datasets was that all along the globe, a system of mid-oceanic ridges was detected. An important conclusion was that along this system, new ocean floor was being created, which led to the concept of the "Great Global Rift". This was described in the crucial paper of Bruce Heezen (1960), which would trigger a real revolution in thinking. A profound consequence of seafloor spreading is that new crust was, and still is, being continually created along the oceanic ridges. Therefore, Heezen advocated the so-called "expanding Earth" hypothesis of S. Warren Carey. So, still the question remained: how can new crust be continuously added along the oceanic ridges without increasing the size of the Earth? In reality, this question had been solved already by numerous scientists during the forties and the fifties, like Arthur Holmes, Vening-Meinesz, Coates and many others: The crust in excess disappeared along what were called the oceanic trenches, where so-called "subduction" occurred. Therefore, when various scientists during the early sixties started to reason on the data at their disposal regarding the ocean floor, the pieces of the theory quickly fell into place.

The question particularly intrigued Harry Hammond Hess, a Princeton University geologist and a Naval Reserve Rear Admiral, and Robert S. Dietz, a scientist with the U. S. Coast and Geodetic Survey who first coined the term seafloor spreading. Dietz and Hess (the former published the same idea one year earlier in Nature, but priority belongs to Hess who had already distributed an unpublished manuscript of his 1962 article by 1960) were among the small handful who really understood the broad implications of sea floor spreading and how it would eventually agree with the, at that time, unconventional and unaccepted ideas of continental drift and the elegant and mobilistic models proposed by previous workers like Holmes.

In the same year, Robert R. Coats of the U. S. Geological Survey described the main features of island arc subduction in the Aleutian Islands. His paper, though little noted (and even ridiculed) at the time, has since been called "seminal" and "prescient". In reality, it actually shows that the work by the European scientists on island arcs and mountain belts performed and published during the 1930s up until the 1950s was applied and appreciated also in the United States.

If the Earth's crust was expanding along the oceanic ridges, Hess and Dietz reasoned like Holmes and others before them, it must be shrinking elsewhere. Hess followed Heezen, suggesting that new oceanic crust continuously spreads away from the ridges in a conveyor belt–like motion. And, using the mobilistic concepts developed before, he correctly concluded that many millions of years later, the oceanic crust eventually descends along the continental margins where oceanic trenches – very deep, narrow canyons – are formed, e. g. along the rim of the Pacific Ocean basin. The important step Hess made was that convection currents would be the driving force in this process, arriving at the same conclusions as Holmes had decades before with the only difference that the thinning of the ocean crust was performed using Heezen's mechanism of spreading along the ridges. Hess therefore concluded that the Atlantic Ocean was expanding while the Pacific Ocean was shrinking. As old oceanic crust is "consumed" in the trenches (like Holmes and others, he thought this was done by thickening of the continental lithosphere, not, as now understood, by underthrusting at a larger scale of the oceanic crust itself into the mantle), new magma rises and erupts along the spreading ridges to form new crust. In effect, the ocean basins are perpetually being "recycled, " with the creation of new crust and the destruction of old oceanic lithosphere occurring simultaneously. Thus, the new mobilistic concepts neatly explained why the Earth does not get bigger with sea floor spreading, why there is so little sediment accumulation on the ocean floor, and why oceanic rocks are much younger than continental rocks.

Magnetic Striping

Beginning in the 1950s, scientists like Victor Vacquier, using magnetic instruments (magnetometers) adapted from airborne devices developed during World War II to detect submarines, began recognizing odd magnetic variations across the ocean floor. This finding, though unexpected, was not entirely surprising because it was known that basalt—the iron-rich, volcanic rock making up the ocean floor—contains a strongly magnetic mineral (magnetite) and can locally distort compass readings. This distortion was recognized by Icelandic mariners as early as the late 18th century. More important, because the presence of magnetite gives the basalt measurable magnetic properties, these newly discovered magnetic variations provided another means to study the deep ocean floor. When newly formed rock cools, such magnetic materials recorded the Earth's magnetic field at the time.

As more and more of the seafloor was mapped during the 1950s, the magnetic variations turned out not to be random or isolated occurrences, but instead revealed recognizable patterns. When these magnetic patterns were mapped over a wide region, the ocean floor showed a zebra-like pattern: one stripe with normal polarity and the adjoining stripe with reversed polarity. The overall pattern, defined by these alternating bands of normally and reversely polarized rock, became known as magnetic striping, and was published by Ron G. Mason and co-workers in 1961, who did not find, though, an explanation for these data in terms of sea floor spreading, like Vine, Matthews and Morley a few years later.

The discovery of magnetic striping called for an explanation. In the early 1960s scientists such as Heezen, Hess and Dietz had begun to theorise that mid-ocean ridges mark structurally weak zones where the ocean floor was being ripped in two lengthwise along the ridge crest.

New magmafrom deep within the Earth rises easily through these weak zones and eventually erupts along the crest of the ridges to create new oceanic crust. This process, at first denominated the "conveyer belt hypothesis" and later called seafloor spreading, operating over many millions of years continues to form new ocean floor all across the 50, 000 km-long system of mid-ocean ridges.

Only four years after the maps with the "zebra pattern" of magnetic stripes were published, the link between sea floor spreading and these patterns was correctly placed, independently by Lawrence Morley, and by Fred Vine and Drummond Matthews, in 1963, now called theVine-Matthews-Morley hypothesis. This hypothesis linked these patterns to geomagnetic reversals and was supported by several lines of evidence:

1. the stripes are symmetrical around the crests of the mid-ocean ridges; at or near the crest of the ridge, the rocks are very young, and they become progressively older away from the ridge crest;

2. the youngest rocks at the ridge crest always have present-day (normal) polarity;

3. stripes of rock parallel to the ridge crest alternate in magnetic polarity (normal-reversed-normal, etc.), suggesting that they were formed during different epochs documenting the (already known from independent studies) normal and reversal episodes of the Earth's magnetic field.

By explaining both the zebra-like magnetic striping and the construction of the mid-ocean ridge system, the seafloor spreading hypothesis (SFS) quickly gained converts and represented another major advance in the development of the plate-tectonics theory.

Furthermore, the oceanic crust now came to be appreciated as a natural "tape recording" of the history of the geomagnetic field reversals (GMFR) of the Earth's magnetic field. Today, extensive studies are dedicated to the calibration of the normal-reversal patterns in the oceanic crust on one hand and known timescales derived from the dating of basalt layers in sedimentary sequences (magnetostratigraphy) on the other, to arrive at estimates of past spreading rates and plate reconstructions.

Definition and Refining of the Theory

After all these considerations, Plate Tectonics (or, as it was initially called "New

Global Tectonics") became quickly accepted in the scientific world, and numerous papers followed that defined the concepts:

- ◉ In 1965, Tuzo Wilson who had been a promotor of the sea floor spreading hypothesis and continental drift from the very beginningadded the concept of transform faults to the model, completing the classes of fault types necessary to make the mobility of the plates on the globe work out.

- ◉ A symposium on continental drift was held at the Royal Society of London in 1965 which must be regarded as the official start of the acceptance of plate tectonics by the scientific community, and which abstracts are issued as Blacket, Bullard & Runcorn (1965). In this symposium, Edward Bullard and co-workers showed with a computer calculation how the continents along both sides of the Atlantic would best fit to close the ocean, which became known as the famous "Bullard's Fit".

- ◉ In 1966 Wilson published the paper that referred to previous plate tectonic reconstructions, introducing the concept of what is now known as the "Wilson Cycle".

- ◉ In 1967, at the American Geophysical Union's meeting, W. Jason Morgan proposed that the Earth's surface consists of 12 rigid plates that move relative to each other.

- ◉ Two months later, Xavier Le Pichon published a complete model based on 6 major plates with their relative motions, which marked the final acceptance by the scientific community of plate tectonics.

- ◉ In the same year, McKenzie and Parker independently presented a model similar to Morgan's using translations and rotations on a sphere to define the plate motions.

Implications for biogeography

Continental drift theory helps biogeographers to explain the disjunct biogeographic distribution of present day life found on different continents but having similar ancestors. In particular, it explains the Gondwanan distribution of ratites and the Antarctic flora.

Plate reconstruction

Reconstruction is used to establish past (and future) plate configurations, helping determine the shape and make-up of ancient supercontinents and providing a basis for paleogeography.

Defining plate boundaries

Current plate boundaries are defined by their seismicity. Past plate boundaries within existing plates are identified from a variety of evidence, such as the presence of ophiolites that are indicative of vanished oceans.

Past plate motions

Various types of quantitative and semi-quantitative information are available to constrain past plate motions. The geometric fit between continents, such as between west Africa and South America is still an important part of plate reconstruction. Magnetic stripe patterns provide a reliable guide to relative plate motions going back into the Jurassic period. The tracks of hotspots give absolute reconstructions, but these are only available back to the Cretaceous. Older reconstructions rely mainly on paleomagnetic pole data, although these only constrain the latitude and rotation, but not the longitude. Combining poles of different ages in a particular plate to produce apparent polar wander paths provides a method for comparing the motions of different plates through time. Additional evidence comes from the distribution of certain sedimentary rock types, faunal provinces shown by particular fossil groups, and the position of orogenic belts.

FORMATION AND BREAK-UP OF CONTINENTS

The movement of plates has caused the formation and break-up of continents over time, including occasional formation of a supercontinentthat contains most or all of the continents. The supercontinent Columbia or Nuna formed during a period of 2, 000 to 1, 800 million years agoand broke up about 1, 500 to 1, 300 million years ago. The supercontinent Rodinia is thought to have formed about 1 billion years ago and to have embodied most or all of Earth's continents, and broken up into eight continents around 600 million years ago. The eight continents later re-assembled into another supercontinent called Pangaea; Pangaea broke up into Laurasia (which became North America and Eurasia) and Gondwana (which became the remaining continents).

The Himalayas, the world's tallest mountain range, are assumed to have been formed by the collision of two major plates. Before uplift, they were covered by the Tethys Ocean.

Gallery of past configurations

Interpretive simulation of past continental movement and shorelines, with time given in millions of years ago (Ma). For more complete timeline of images, see Gallery of continental movement.

Current plates

Depending on how they are defined, there are usually seven or eight "major" plates: African, Antarctic, Eurasian, North American, South American, Pacific, and Indo-Australian. The latter is sometimes subdivided into the Indian and Australianplates.

There are dozens of smaller plates, the seven largest of which are the Arabian, Caribbean, Juan de Fuca, Cocos, Nazca, Philippine Sea and Scotia.

The current motion of the tectonic plates is today determined by remote sensing satellite data sets, calibrated with ground station measurements.

The appearance of plate tectonics on terrestrial planets is related to planetary mass, with more massive planets than Earth expected to exhibit plate tectonics. Earth may be a borderline case, owing its tectonic activity to abundant water (silica and water form a deep eutectic.)

Venus

Venus shows no evidence of active plate tectonics. There is debatable evidence of active tectonics in the planet's distant past; however, events taking place since then (such as the plausible and generally accepted hypothesis that the Venusian lithosphere has thickened greatly over the course of several hundred million years) has made constraining the course of its geologic record difficult. However, the numerous well-preserved impact craters have been utilized as a dating method to approximately date the Venusian surface (since there are thus far no known samples of Venusian rock to be dated by more reliable methods). Dates derived are dominantly in the range 500 to 750 million years ago, although ages of up to 1, 200 million years ago have been calculated. This research has led to the fairly well accepted hypothesis that Venus has undergone an essentially complete volcanic resurfacing at least once in its distant past, with the last event taking place approximately within the range of estimated surface ages. While the mechanism of such an impressive thermal event remains a debated issue in Venusian geosciences, some scientists are advocates of processes involving plate motion to some extent.

One explanation for Venus' lack of plate tectonics is that on Venus temperatures are too high for significant water to be present. The Earth's crust is soaked with water, and water plays an important role in the development of shear zones. Plate tectonics requires weak surfaces in the crust along which crustal slices can move, and it may well be that such weakening never took place on Venus because of the absence of water. However, some researchers remain convinced that plate tectonics is or was once active on this planet.

Mars

Mars is considerably smaller than Earth and Venus, and there is evidence for ice on its surface and in its crust.

In the 1990s, it was proposed that Martian Crustal Dichotomy was created by plate tectonic processes. Scientists today disagree, and believe that it was created either by upwelling within the Martian mantle that thickened the crust of the Southern Highlands and formedTharsis or by a giant impact that excavated the Northern Lowlands.

Valles Marineris may be a tectonic boundary. Observations made of the magnetic field of Mars by the Mars Global Surveyor spacecraft in 1999 showed patterns of magnetic striping discovered on this planet. Some scientists interpreted these as requiring plate tectonic processes, such as seafloor spreading. However, their data fail a "magnetic reversal test", which is used to see if they were formed by flipping polarities of a global magnetic field.

Galilean satellites of Jupiter

Some of the satellites of Jupiter have features that may be related to plate-tectonic style deformation, although the materials and specific mechanisms may be different from plate-tectonic activity on Earth. On 8 September 2014, NASA reported finding evidence of plate tectonics on Europa, a satellite of Jupiter - the first sign of such geological activity on another world other than Earth.

Titan, moon of Saturn

Titan, the largest moon of Saturn, was reported to show tectonic activity in images taken by the Huygens Probe, which landed on Titan on January 14, 2005.

Exoplanets

On Earth-sized planets, plate tectonics is more likely if there are oceans of water; however, in 2007, two independent teams of researchers came to opposing conclusions about the likelihood of plate tectonics on larger super-earths with one team saying that plate tectonics would be episodic or stagnant and the other team saying that plate tectonics is very likely on super-earths even if the planet is dry.

GROUND INVESTIGATIONS

14

Geotechnical investigations are performed by geotechnical engineers or engineering geologists to obtain information on the physical properties of soil and rock around a site to design earthworks and foundations for proposed structures and for repair of distress to earthworks and structures caused by subsurface conditions. This type of investigation is called a site investigation. Additionally, geotechnical investigations are also used to measure thethermal resistivity of soils or backfill materials required for underground transmission lines, oil and gas pipelines, radioactive waste disposal, and solar thermal storage facilities. A geotechnical investigation will include surface exploration and subsurface exploration of a site. Sometimes, geophysical methods are used to obtain data about sites. Subsurface exploration usually involves soil sampling and laboratory tests of the soil samples retrieved.

Surface exploration can include geologic mapping, geophysical methods, and photogrammetry, or it can be as simple as a geotechnical professional walking around on the site to observe the physical conditions at the site.

To obtain information about the soil conditions below the surface, some form of subsurface exploration is required. Methods of observing the soils below the surface, obtaining samples, and determining physical properties of the soils and rocks include test pits, trenching (particularly for locating faults and slide planes), boring, and in situ tests.

SOIL SAMPLING

Borings come in two main varieties, large-diameter and small-diameter. Large-diameter borings are rarely used due to safety concerns and expense, but are sometimes used to allow a geologist or engineer to visually and manually examine the soil and rock stratigraphy in-situ. Small-diameter borings are frequently used to allow a geologist or engineer to examine soil or rock cuttings or to retrieve samples at depth using soil samplers, and to perform in-place soil tests.

Soil samples are often categorized as being either "disturbed" or "undisturbed;" however, "undisturbed" samples are not truly undisturbed. A disturbed sample is one in which the structure of the soil has been changed sufficiently that tests of structural properties of the soil will not be representative of in-situ conditions, and only properties of the soil grains (e. g., grain size distribution, Atterberg limits, and possibly the water content) can be accurately determined. An undisturbed sample is one where the condition of the soil in the sample is close enough to the conditions of the soil in-situ to allow tests of structural properties of the soil to be used to approximate the properties of the soil in-situ.

Offshore soil collection introduces many difficult variables. In shallow water, work can be done off a barge. In deeper water a ship will be required. Deepwater soil samplers are normally variants of Kullenberg-type samplers, a modification on a basic gravity corer using a piston (Lunne and Long, 2006). Seabed samplers are also available, which push the collection tube slowly into the soil.

Soil Samplers

Soil samples are taken using a variety of samplers; some provide only disturbed samples, while others can provide relatively undisturbed samples.

- ◉ Shovel. Samples can be obtained by digging out soil from the site. Samples taken this way are disturbed samples.

- ◉ Trial Pits are relatively small hand or machine excavated tranches used to determine groundwater levels and take disturbed samples from.

- ◉ Hand/Machine Driven Auger. This sampler typically consists of a short cylinder with a cutting edge attached to a rod and handle. The sampler is advanced by a combination of rotation and downward force. Samples taken this way are disturbed samples.

- ◉ Continuous Flight Auger. A method of sampling using an auger as a corkscrew. The auger is screwed into the ground then lifted out. Soil is retained on the blades of the auger and kept for testing. The soil sampled this way is considered disturbed.

- ◉ Split-spoon / SPT Sampler. Utilized in the 'Standard Test Method for Standard Penetration Test (SPT) and Split-Barrel Sampling of Soils' (ASTM D 1586). This sampler is typically an 18"-30" long, 2. 0" outside diameter (OD) hollow tube split in half lengthwise. A hardened metal drive shoe with a 1. 375" opening is attached to the bottom end, and a one-way valve and drill rod adapter at the sampler head. It is driven into the ground with a 140-pound (64 kg) hammer falling 30". The blow counts (hammer strikes) required to advance the sampler a total of 18" are counted and reported. Generally used for non-cohesive soils, samples taken this way are considered disturbed.

- Modified California Sampler. Utilized in the 'Standard Practice for Thick Wall, Ring-Lined, Split Barrel, Drive Sampling of Soils1' (ASTM D 3550). Similar in concept to the SPT sampler, the sampler barrel has a larger diameter and is usually lined with metal tubes to contain samples. Samples from the Modified California Sampler are considered disturbed due to the large area ratio of the sampler (sampler wall area/sample cross sectional area).

- Shelby Tube Sampler. Utilized in the 'Standard Practice for Thin-Walled Tube Sampling of Soils for Geotechnical Purposes' (ASTM D 1587). This sampler consists of a thin-walled tube with a cutting edge at the toe. A sampler head attaches the tube to the drill rod, and contains a check valve and pressure vents. Generally used in cohesive soils, this sampler is advanced into the soil layer, generally 6" less than the length of the tube. The vacuum created by the check valve and cohesion of the sample in the tube cause the sample to be retained when the tube is withdrawn. Standard ASTM dimensions are; 2" OD, 36" long, 18 gauge thickness; 3" OD, 36" long, 16 gauge thickness; and 5" OD, 54" long, 11 gauge thickness. It should be noted that ASTM allows other diameters as long as they are proportional to the standardized tube designs, and tube length is to be suited for field conditions. Soil sampled in this manner is considered undisturbed.

- Piston samplers. These samplers are thin-walled metal tubes which contain a piston at the tip. The samplers are pushed into the bottom of a borehole, with the piston remaining at the surface of the soil while the tube slides past it. These samplers will return undisturbed samples in soft soils, but are difficult to advance in sands and stiff clays, and can be damaged (compromising the sample) if gravel is encountered. The Livingstone corer, developed by D. A. Livingstone, is a commonly used piston sampler. A modification of the Livingstone corer with a serrated coring head allows it to be rotated to cut through subsurface vegetable matter such as small roots or buried twigs.

- Pitcher Barrel sampler. This sampler is similar to piston samplers, except that there is no piston. There are pressure-relief holes near the top of the sampler to prevent pressure buildup of water or air above the soil sample.

In-situ tests

- A standard penetration test is an in-situ dynamic penetration test designed to provide information on the properties of soil, while also collecting a disturbed soil sample for grain-size analysis and soil classification.

- A dynamic cone penetrometer is an insitu test in which a weight is manually lifted and dropped on a cone which penetrates the ground. the number of mm per hit are recorded and this is used to estimate certain soil properties.

This is a simple test method and usually needs backing up with lab data to get a good correlation.

⊙ A cone penetration test is performed using an instrumented probe with a conical tip, pushed into the soil hydraulically at a constant rate. A basic CPT instrument reports tip resistance and shear resistance along the cylindrical barrel. CPT data has been correlated to soil properties. Sometimes instruments other than the basic CPT probe are used, including:

⊙ A piezocone penetrometer probe is advanced using the same equipment as a regular CPT probe, but the probe has an additional instrument which measures the groundwater pressure as the probe is advanced.

⊙ A seismic piezocone penetrometer probe is advanced using the same equipment as a CPT or CPTu probe, but the probe is also equipped with either geophones or accelerometers to detect shear waves and/or pressure waves produced by a source at the surface.

⊙ Full flow penetrometers (T-bar, ball, and plate) probes are used in extremely soft clay soils (such as sea-floor deposits) and are advanced in the same manner as the CPT. As their names imply, the T-bar is a cylindrical bar attached at right angles to the drill string forming what look likes a T, the ball is a large sphere, and the plate is flat circular plate. In soft clays, soil flows around the probe similar to a viscous fluid. The pressure due to overburden stress and pore water pressure is equal on all sides of the probes (unlike with CPT's), so no correction is necessary, reducing a source of error and increasing accuracy. Especially desired in soft soils due to the very low loads on the measuring sensors. Full flow probes can also be cycled up and down to measure remolded soil resistance. Ultimately the geotechnical professional can use the measured penetration resistance to estimate undrained and remolded shear strengths.

⊙ Helical probe test soil exploration and compaction testing by the helical probe test (HPT) has become popular for providing a quick and accurate method of determining soil properties at relatively shallow depths. The HPT test is attractive for in-situ footing inspections because it is lightweight and can be conducted quickly by one person. During testing, the probe is driven to the desired depth and the torque required to turn the probe is used as a measure to determine the soil's characteristics. Preliminary ASTM testing has determined that the HPT method correlates well to standard penetration testing (SPT) and cone penetration testing (CPT) with empirical calibration.

A flat plate dilatometer test (DMT) is a flat plate probe often advanced using CPT rigs, but can also be advanced from conventional drill rigs. A diaphragm on the plate applies a lateral force to the soil materials and measures the strain induced for various levels of applied stress at the desired depth interval.

In-situ gas tests can be carried out in the boreholes on completion and in probe holes made in the sides of the trial pits as part of the site investigation. Testing is normally with a portable meter, which measures the methane content as its percentage volume in air. The corresponding oxygen and carbon dioxide concentrations are also measured. A more accurate method used to monitor over the longer term, consists of gas monitoring standpipes should be installed in boreholes.

These typically comprise slotted uPVC pipework surrounded by single sized gravel. The top 0. 5 m to 1. 0 m of pipework is usually not slotted and is surrounded by bentonite pellets to seal the borehole. Valves are fitted and the installations protected by lockable stopcock covers normally fitted flush with the ground. Monitoring is again with a portable meter and is usually done on a fortnightly or monthly basis.

LABORATORY TESTS

A wide variety of laboratory tests can be performed on soils to measure a wide variety of soil properties. Some soil properties are intrinsic to the composition of the soil matrix and are not affected by sample disturbance, while other properties depend on the structure of the soil as well as its composition, and can only be effectively tested on relatively undisturbed samples.

Some soil tests measure direct properties of the soil, while others measure "index properties" which provide useful information about the soil without directly measuring the property desired.

Atterberg Limits

The Atterberg limits define the boundaries of several states of consistency for plastic soils. The boundaries are defined by the amount of water a soil needs to be at one of those boundaries.

The boundaries are called the plastic limit and the liquid limit, and the difference between them is called the plasticity index. The shrinkage limit is also a part of the Atterberg limits. The results of this test can be used to help predict other engineering properties.

California Bearing Ratio

ASTM D 1883. A test to determine the aptitude of a soil or aggregate sample as a road subgrade. A plunger is pushed into a compacted sample, and its resistance is measured.

This test was developed by Caltrans, but it is no longer used in the Caltrans pavement design method. It is still used as a cheap method to estimate the resilient modulus.

Direct Shear Test

ASTM D3080. The direct shear test determines the consolidated, drained strength properties of a sample. A constant strain rate is applied to a single shear plane under a normal load, and the load response is measured. If this test is performed with different normal loads, the common shear strength parameters can be determined.

Expansion Index Test

This test uses a remolded soil sample to determine the Expansion Index (EI), an empirical value required by building design codes, at a water content of 50% for expansive soils, like expansive clays.

Hydraulic conductivity tests

There are several tests available to determine a soil's hydraulic conductivity. They include the constant head, falling head, and constant flow methods. The soil samples tested can be any type include remolded, undisturbed, and compacted samples.

Oedometer test

This can be used to determine consolidation (ASTM D2435) and swelling (ASTM D4546) parameters.

Particle-size Analysis

This is done to determine the soil gradation. Coarser particles are separated in the sieve analysis portion, and the finer particles are analyzed with a hydrometer. The distinction between coarse and fine particles is usually made at 75 ìm. The sieve analysis shakes the sample through progressively smaller meshes to determine its gradation. The hydrometer analysis uses the rate of sedimentation to determine particle gradation.

R-Value Test

California Test 301 This test measures the lateral response of a compacted sample of soil or aggregate to a vertically applied pressure under specific conditions. This test is used by Caltrans for pavement design, replacing the California bearing ratio test.

Soil compaction Tests

Standard Proctor (ASTM D698), Modified Proctor (ASTM D1557), and California Test 216. These tests are used to determine the maximum unit weight and optimal water content a soil can achieve for a given compaction effort.

Soil suction tests

ASTM D5298.

Triaxial shear tests

This is a type of test that is used to determine the shear strength properties of a soil. It can simulate the confining pressure a soil would see deep into the ground. It can also simulate drained and undrained conditions.

Unconfined compression test

ASTM D2166. This test compresses a soil sample to measure its strength. The modifier "unconfined" contrasts this test to the triaxial shear test.

Water content

This test provides the water content of the soil, normally expressed as a percentage of the weight of water to the dry weight of the soil.

Geophysical exploration

Geophysical methods are used in geotechnical investigations to evaluate a site's behavior in a seismic event. By measuring a soil's shear wave velocity, the dynamic response of that soil can be estimated. There are a number of methods used to determine a site's shear wave velocity:

- Crosshole method
- Downhole method (with a seismic CPT or a substitute device)
- Surface wave reflection or refraction
- Suspension logging (also known as P-S logging or Oyo logging)
- Spectral analysis of surface waves (SASW)
- Multichannel analysis of surface waves (MASW)
- Reflection microtremor (ReMi)

Other methods:

- Electromagnetic (radar, resistivity)
- Optical/acoustic televiewer survey

SOIL STRENGTH 15

Soil strength is a term used in soil mechanics to describe the magnitude of the shear stress that a soil can sustain. The shear resistance of soil is a result of friction and interlocking of particles, and possibly cementation or bonding at particle contacts. Due to interlocking, particulate material may expand or contract in volume as it is subject to shear strains. If soil expands its volume, the density of particles will decrease and the strength will decrease; in this case, the peak strength would be followed by a reduction of shear stress. The stress-strain relationship levels off when the material stops expanding or contracting, and when interparticle bonds are broken. The theoretical state at which the shear stress and density remain constant while the shear strain increases may be called the critical state, steady state, or residual strength.

The volume change behavior and interparticle friction depend on the density of the particles, the intergranular contact forces, and to a somewhat lesser extent, other factors such as the rate of shearing and the direction of the shear stress. The average normal intergranular contact force per unit area is called the effective stress.

If water is not allowed to flow in or out of the soil, the stress path is called an undrained stress path. During undrained shear, if the particles are surrounded by a nearly incompressible fluid such as water, then the density of the particles cannot change without drainage, but the water pressure and effective stress will change. On the other hand, if the fluids are allowed to freely drain out of the pores, then the pore pressures will remain constant and the test path is called a drained stress path. The soil is free to dilate or contract during shear if the soil is drained. In reality, soil is partially drained, somewhere between the perfectly undrained and drained idealized conditions. The shear strength of soil depends on the effective stress, the drainage conditions, the density of the particles, the rate of strain, and the direction of the strain.

For undrained, constant volume shearing, the Tresca theory may be used to predict the shear strength, but for drained conditions, the Mohr–Coulomb theory may be used.

Two important theories of soil shear are the critical state theory and the steady state theory. There are key differences between the critical state condition and the steady state condition and the resulting theory corresponding to each of these conditions.

Factors Controlling Shear Strength of Soils

The stress-strain relationship of soils, and therefore the shearing strength, is affected (Poulos 1989) by:

- Soil composition (basic soil material): mineralogy, grain size and grain size distribution, shape of particles, pore fluid type and content, ions on grain and in pore fluid.

- State (initial): Defined by the initial void ratio, effective normal stress and shear stress (stress history). State can be described by terms such as: loose, dense, overconsolidated, normally consolidated, stiff, soft, contractive, dilative, etc.

- Structure: Refers to the arrangement of particles within the soil mass; the manner the particles are packed or distributed. Features such as layers, joints, fissures, slickensides, voids, pockets, cementation, etc., are part of the structure. Structure of soils is described by terms such as: undisturbed, disturbed, remolded, compacted, cemented; flocculent, honey-combed, single-grained; flocculated, deflocculated; stratified, layered, laminated; isotropic and anisotropic.

- Loading conditions: Effective stress path, i. e., drained, and undrained; and type of loading, i. e., magnitude, rate (static, dynamic), and time history (monotonic, cyclic).

Undrained strength

This term describes a type of shear strength in soil mechanics as distinct from drained strength.

Conceptually, there is no such thing as the undrained strength of a soil. It depends on a number of factors, the main ones being:

- Orientation of stresses
- Stress path
- Rate of shearing
- Volume of material (like for fissured clays or rock mass)

Undrained strength is typically defined by Tresca theory, based on Mohr's circle as:

$$s1 - s3 = 2\ Su$$

Where:

- ◉ s1 is the major principal stress
- ◉ s3 is the minor principal stress
- ◉ is the shear strength (s1 - s3)/2
- ◉ hence, = Su (or sometimes cu), the undrained strength.

It is commonly adopted in limit equilibrium analyses where the rate of loading is very much greater than the rate at which pore water pressures, that are generated due to the action of shearing the soil, may dissipate. An example of this is rapid loading of sands during an earthquake, or the failure of a clay slope during heavy rain, and applies to most failures that occur during construction.

As an implication of undrained condition, no elastic volumetric strains occur, and thus Poisson's ratio is assumed to remain 0. 5 throughout shearing. The Tresca soil model also assumes no plastic volumetric strains occur. This is of significance in more advanced analyses such as infinite element analysis. In these advanced analysis methods, soil models other than Tresca may be used to model the undrained condition including Mohr-Coulomb and critical state soil models such as the modified Cam-clay model, provided Poisson's ratio is maintained at 0. 5.

One relationship used extensively by practicing engineers is the empirical observation that the ratio of the undrained shear strength c to the original consolidation stress p' is approximately a constant for a given Over Consolidation Ratio (OCR). This relationship was first formalized by (Henkel 1960) and (Henkel & Wade 1966) who also extended it to show that stress-strain characteristics of remolded clays could also be normalized with respect to the original consolidation stress. The constant c/p relationship can also be derived from theory for both critical-state and steady-state soil mechanics (Joseph 2012). This fundamental, normalization property of the stress-strain curves is found in many clays, and was refined into the empirical SHANSEP (stress history and normalized soil engineering properties) method.

Drained shear strength

The drained shear strength is the shear strength of the soil when pore fluid pressures, generated during the course of shearing the soil, are able to dissipate during shearing. It also applies where no pore water exists in the soil (the soil is dry) and hence pore fluid pressures are negligible. It is commonly approximated using the Mohr-Coulomb equation. (It was called "Coulomb's equation" by Karl von Terzaghi in 1942.) (Terzaghi 1942) combined it with the principle of effective stress.

In terms of effective stresses, the shear strength is often approximated by:

$$= s' \tan(j') + c'$$

Where $s' = (s - u)$, is defined as the effective stress. s is the total stress applied normal to the shear plane, and u is the pore water pressure acting on the same plane.

j' = the effective stress friction angle, or the 'angle of internal friction' after Coulomb friction. The coefficient of friction is equal to $\tan(j')$. Different values of friction angle can be defined, including the peak friction angle, j'_p, the critical state friction angle, j'_{cv}, or residual friction angle, j'_r.

c' = is called cohesion, however, it usually arises as a consequence of forcing a straight line to fit through measured values of $(ô, s')$ even though the data actually falls on a curve. The intercept of the straight line on the shear stress axis is called the cohesion. It is well known that the resulting intercept depends on the range of stresses considered: it is not a fundamental soil property. The curvature (nonlinearity) of the failure envelope occurs because the dilatancy of closely packed soil particles depends on confining pressure.

Critical state Theory

A more advanced understanding of the behaviour of soil undergoing shearing lead to the development of the critical state theory of soil mechanics (Roscoe, Schofield & Wroth 1958). In critical state soil mechanics, a distinct shear strength is identified where the soil undergoing shear does so at a constant volume, also called the 'critical state'. Thus there are three commonly identified shear strengths for a soil undergoing shear:

- ◉ Peak strength p

- ◉ Critical state or constant volume strength cv

- ◉ Residual strength r

The peak strength may occur before or at critical state, depending on the initial state of the soil particles being sheared:

- ◉ A loose soil will contract in volume on shearing, and may not develop any peak strength above critical state. In this case 'peak' strength will coincide with the critical state shear strength, once the soil has ceased contracting in volume. It may be stated that such soils do not exhibit a distinct 'peak strength'.

- ◉ A dense soil may contract slightly before granular interlock prevents further contraction (granular interlock is dependent on the shape of the grains and their initial packing arrangement). In order to continue shearing once granular interlock has occurred, the soil must dilate (expand in volume). As additional shear force is required to dilate the soil, a 'peak' strength occurs. Once this peak strength caused by dilation has been overcome through continued shearing, the resistance provided by the soil to the applied shear stress reduces (termed "strain softening"). Strain softening will continue until no further changes in volume of the soil occur on continued shearing. Peak strengths are also observed in overconsolidated clays where the natural fabric of the soil must be destroyed prior to reaching constant volume shearing. Other effects that result in peak strengths include cementation and bonding of particles.

The constant volume (or critical state) shear strength is said to be intrinsic to the soil, and independent of the initial density or packing arrangement of the soil grains. In this state the grains being sheared are said to be 'tumbling' over one another, with no significant granular interlock or sliding plane development affecting the resistance to shearing. At this point, no inherited fabric or bonding of the soil grains affects the soil strength.

The residual strength occurs for some soils where the shape of the particles that make up the soil become aligned during shearing (forming aslickenside), resulting in reduced resistance to continued shearing (further strain softening). This is particularly true for most clays that comprise plate-like minerals, but is also observed in some granular soils with more elongate shaped grains. Clays that do not have plate-like minerals (like allophanic clays) do not tend to exhibit residual strengths.

Use in practice: If one is to adopt critical state theory and take c' = 0; p may be used, provided the level of anticipated strains are taken into account, and the effects of potential rupture or strain softening to critical state strengths are considered. For large strain deformation, the potential to form slickensided surface with a j'r should be considered (such as pile driving).

The Critical State occurs at the quasi-static strain rate. It does not allow for differences in shear strength based on different strain rates. Also at the critical state, there is no particle alignment or specific soil structure.

Steady state (dynamical systems based soil shear)

A refinement of the critical state concept is the steady state concept.

The steady state strength is defined as the shear strength of the soil when it is at the steady state condition. The steady state condition is defined (Poulos 1981) as "that state in which the mass is continuously deforming at constant volume, constant normal effective stress, constant shear stress, and constant velocity." Steve J. Poulos, then an Associate Professor of the Soil Mechanics Department of Harvard

University, built off a hypothesis that Arthur Casagrande was formulating towards the end of his career. (Poulos 1981) Steady state based soil mechanics is sometimes called "Harvard soil mechanics". The steady state condition is not the same as the "critical state" condition.

The steady state occurs only after all particle breakage if any is complete and all the particles are oriented in a statistically steady state condition and so that the shear stress needed to continue deformation at a constant velocity of deformation does not change. It applies to both the drained and the undrained case.

The steady state has a slightly different value depending on the strain rate at which it is measured. Thus the steady state shear strength at the quasi-static strain rate (the strain rate at which the critical state is defined to occur at) would seem to correspond to the critical state shear strength. However there is an additional difference between the two states. This is that at the steady state condition the grains position themselves in the steady state structure, whereas no such structure occurs for the critical state. In the case of shearing to large strains for soils with elongated particles, this steady state structure is one where the grains are oriented (perhaps even aligned) in the direction of shear. In the case where the particles are strongly aligned in the direction of shear, the steady state corresponds to the "residual condition. "

Three common misconceptions regarding the steady state are that a) it is the same as the critical state (it is not), b) that it applies only to the undrained case (it applies to all forms of drainage), and c) that it does not apply to sands (it applies to any granular material). A primer on the Steady State theory can be found in a report by Poulos (Poulos 1971). Its use in earthquake engineering is described in detail in another publication by Poulos (Poulos 1989).

The difference between the steady state and the critical state is not merely one of semantics as is sometimes thought, and it is incorrect to use the two terms/concepts interchangeably. The additional requirements of the strict definition of the steady state over and above the critical state viz. a constant deformation velocity and statistically constant structure (the steady state structure), places the steady state condition within the framework of dynamical systems theory. This strict definition of the steady state was used to describe soil shear as a dynamical system (Joseph 2009). Dynamical systems are ubiquitous in nature (the Great Red Spot on Jupiter is one example) and mathematicians have extensively studied such systems. The underlying basis of the soil shear dynamical system is simple friction (Joseph 2012).

CHAPTER

SLOPE FAILURE AND LANDSLIDES 16

Slope stability is the potential of soil covered slopes to withstand and undergo movement. Stability is determined by the balance of shear stress and shear strength. A previously stable slope may be initially affected by preparatory factors, making the slope conditionally unstable. Triggering factors of a slope failure can be climatic events can then make a slope actively unstable, leading to mass movements. Mass movements can be caused by increase in shear stress, such as loading, lateral pressure, and transient forces. Alternatively, shear strength may be decreased by weathering, changes in pore water pressure, and organic material.

The field of slope stability encompasses static and dynamic stability of slopes of earth and rock-fill dams, slopes of other types of embankments, excavated slopes, and natural slopes in soil and soft rock. Slope stability investigation, analysis (including modeling), and design mitigation is typically completed by geologists, engineering geologists, or geotechnical engineers. Geologists and engineering geologists can also use their knowledge of earth process and their ability to interpret surface geomorphology to determine relative slope stability based simply on site observations.

EXAMPLES

As seen in Figure 1, earthen slopes can develop a cut-spherical weakness area. The probability of this happening can be calculated in advance using a simple 2-D circular analysis package. A primary difficulty with analysis is locating the most-probable slip plane for any given situation. Many landslides have only been analyzed after the fact. More recently slope stability radartechnology has been employed, particularly in the mining industry, to gather real time data and assist in pro-actively determining the likelihood of slope failure.

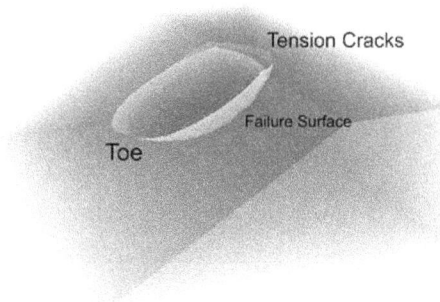

Figure 1: Simple slope slip section

Real life failures in naturally deposited mixed soils are not necessarily circular, but prior to computers, it was far easier to analyse such a simplified geometry. Nevertheless, failures in 'pure' clay can be quite close to circular. Such slips often occur after a period of heavy rain, when the pore water pressure at the slip surface increases, reducing the effective normal stress and thus diminishing the restraining friction along the slip line. This is combined with increased soil weight due to the added groundwater. A 'shrinkage' crack (formed during prior dry weather) at the top of the slip may also fill with rain water, pushing the slip forward. At the other extreme, slab-shaped slips on hillsides can remove a layer of soil from the top of the underlying bedrock. Again, this is usually initiated by heavy rain, sometimes combined with increased loading from new buildings or removal of support at the toe (resulting from road widening or other construction work). Stability can thus be significantly improved by installing drainage paths to reduce the destabilising forces. Once the slip has occurred, however, a weakness along the slip circle remains, which may then recur at the next monsoon.

Slope stability issues can be seen with almost any walk down a ravine in an urban setting. An example is shown in Figure 3, where a river is eroding the toe of a slope, and there is a swimming pool near the top of the slope. If the toe is eroded too far, or the swimming pool begins to leak, the forces driving a slope failure will exceed those resisting failure, and a landslide will develop, possibly quite suddenly.

ANALYSIS METHODS

If the forces available to resist movement are greater than the forces driving movement, the slope is considered stable. A factor of safety is calculated by dividing the forces resisting movement by the forces driving movement. In earthquake-prone areas, the analysis is typically run for static conditions and pseudo-static conditions, where the seismic forces from an earthquake are assumed to add static loads to the analysis.

Landslide

A landslide, also known as a landslip, is a geological phenomenon that includes a wide range of ground movements. Rockfalls, deep failure of slopesand shallow debris flows were common. Landslides can occur in offshore, coastal and onshore environments. Although the action of gravity is the primary driving force for a landslide to occur, there are other contributing factors affecting the original slope stability. Typically, pre-conditional factors build up specific sub-surface conditions that make the area/ slope prone to failure, whereas the actual landslide often requires a trigger before being released.

Causes

Landslides occur when the stability of the slope changes from a stable to an unstable condition. A change in the stability of a slope can be caused by a number of factors, acting together or alone. Natural causes of landslides include:

- Groundwater (pore water) pressure acting to destabilize the slope
- Loss or absence of vertical vegetative structure, soil nutrients, and soil structure (e. g. after a wildfire - a fire in forests lasting for 3–4 days)
- Erosion of the toe of a slope by rivers or ocean waves
- Weakening of a slope through saturation by snow melt, glaciers melting, or heavy rains
- Earthquakes adding loads to barely stable slope
- Earthquake-caused liquefaction destabilizing slopes
- Volcanic eruptions Landslides are aggravated by human activities, such as
- Deforestation, cultivation and construction, which destabilize the already fragile slopes.
- Vibrations from machinery or traffic
- Blasting
- Earthwork which alters the shape of a slope, or which imposes new loads on an existing slope
- In shallow soils, the removal of deep-rooted vegetation that binds colluvium to bedrock
- Construction, agricultural or forestry activities (logging) which change the amount of water which infiltrates the soil.

Types

Debris flow

Slope material that becomes saturated with water may develop into a debris flow or mud flow. The resulting slurry of rock and mud may pick up trees, houses and cars, thus blocking bridges andtributaries causing flooding along its path.

Debris flow is often mistaken for flash flood, but they are entirely different processes.

Muddy-debris flows in alpine areas cause severe damage to structures and infrastructure and often claim human lives. Muddy-debris flows can start as a result of slope-related factors and shallow landslides can dam stream beds, resulting in temporary water blockage. As the impoundments fail, a "domino effect" may be created, with a remarkable growth in the volume of the flowing mass, which takes up the debris in the stream channel. The solid-liquid mixture can reach densities of up to 2 tons/m^3 and velocities of up to 14 m/s (Chiarle and Luino, 1998; Arattano, 2003). These processes normally cause the first severe road interruptions, due not only to deposits accumulated on the road (from several cubic metres to hundreds of cubic metres), but in some cases to the complete removal of bridges or roadways or railways crossing the stream channel. Damage usually derives from a common underestimation of mud-debris flows: in the alpine valleys, for example, bridges are frequently destroyed by the impact force of the flow because their span is usually calculated only for a water discharge. For a small basin in the Italian Alps (area = 1. 76 km^2) affected by a debris flow, Chiarle and Luino (1998) estimated a peak discharge of 750 m3/s for a section located in the middle stretch of the main channel. At the same cross section, the maximum foreseeable water discharge (by HEC-1), was 19 m^3/s, a value about 40 times lower than that calculated for the debris flow that occurred.

Earthflows

Earthflows are downslope, viscous flows of saturated, fine-grained materials, which move at any speed from slow to fast. Typically, they can move at speeds from 0. 17 to 20 km/h (0. 1 to 12. 4 mph). Though these are a lot like mudflows, overall they are more slow moving and are covered with solid material carried along by flow from within. They are different from fluid flows because they are more rapid. Clay, fine sand and silt, and fine-grained, pyroclastic material are all susceptible to earthflows. The velocity of the earthflow is all dependent on how much water content is in the flow itself: if there is more water content in the flow, the higher the velocity will be.

These flows usually begin when the pore pressures in a fine-grained mass increase until enough of the weight of the material is supported by pore water to significantly decrease the internal shearing strength of the material. This thereby creates a bulging lobe which advances with a slow, rolling motion. As these lobes spread out, drainage of the mass increases and the margins dry out, thereby lowering the overall velocity of the flow. This process causes the flow to thicken. The bulbous variety of earthflows are not that spectacular, but they are much more common than their rapid counterparts. They develop a sag at their heads and are usually derived from the slumping at the source.

Earthflows occur much more during periods of high precipitation, which saturates the ground and adds water to the slope content. Fissures develop during the movement of clay-like material which creates the intrusion of water into the earthflows. Water then increases the pore-water pressure and reduces the shearing strength of the material.

DEBRIS LANDSLIDE

A debris slide is a type of slide characterized by the chaotic movement of rocks, soil, and debris mixed with water and/or ice. They are usually triggered by the saturation of thickly vegetated slopes which results in an incoherent mixture of broken timber, smaller vegetation and other debris. Debris avalanches differ from debris slides because their movement is much more rapid. This is usually a result of lower cohesion or higher water content and commonly steeper slopes.

Steep coastal cliffs can be caused by catastrophic debris avalanches. These have been common on the submerged flanks of ocean island volcanos such as the Hawaiian Islands and the Cape VerdeIslands. Another slip of this type was Storegga landslide.

Movement: Debris slides generally start with big rocks that start at the top of the slide and begin to break apart as they slide towards the bottom. This is much slower than a debris avalanche. Debris avalanches are very fast and the entire mass seems to liquefy as it slides down the slope. This is caused by a combination of saturated material, and steep slopes. As the debris moves down the slope it generally follows stream channels leaving a v-shaped scar as it moves down the hill. This differs from the more U-shaped scar of a slump. Debris avalanches can also travel well past the foot of the slope due to their tremendous speed.

Sturzstrom

A sturzstrom is a rare, poorly understood type of landslide, typically with a long run-out. Often very large, these slides are unusually mobile, flowing very far over a low angle, flat, or even slightly uphill terrain.

Shallow landslide

Landslide in which the sliding surface is located within the soil mantle or weathered bedrock (typically to a depth from few decimetres to some metres)is called a shallow landslide. They usually include debris slides, debris flow, and failures of road cut-slopes. Landslides occurring as single large blocks of rock moving slowly down slope are sometimes called block glides.

Shallow landslides can often happen in areas that have slopes with high permeable soils on top of low permeable bottom soils. The low permeable, bottom soils trap the water in the shallower, high permeable soils creating high water pressure in the top soils. As the top soils are filled with water and become heavy, slopes can become very unstable and slide over the low permeable bottom soils. Say there is a slope with silt and sand as its top soil and bedrock as its bottom soil. During an intense rainstorm, the bedrock will keep the rain trapped in the top soils of silt and sand. As the topsoil becomes saturated and heavy, it can start to slide over the bedrock and become a shallow landslide. R. H. Campbell did a study on shallow landslides on Santa Cruz Island California. He notes that if permeability decreases with depth, a perched water table may develop in soils at intense precipitation. When pore water pressures are sufficient to reduce effective normal stress to a critical level, failure occurs.

Deep-seated landslide

Landslides in which the sliding surface is mostly deeply located below the maximum rooting depth of trees (typically to depths greater than ten meters). Deep-seated landslides usually involve deepregolith, weathered rock, and/or bedrock and include large slope failure associated with translational, rotational, or complex movement. This type of landslides are potentially occur in an tectonic active region like Zagros Mountain in Iran. These typically move slowly, only several meters per year, but occasionally move faster. They tend to be larger than shallow landslides and form along a plane of weakness such as a fault or bedding plane. They can be visually identified by concave scarps at the top and steep areas at the toe.

Causing tsunamis

Landslides that occur undersea, or have impact into water, can generate tsunamis. Massive landslides can also generate megatsunamis, which are usually hundreds of meters high. In 1958, one suchtsunami occurred in Lituya Bay in Alaska.

Related phenomena

- An avalanche, similar in mechanism to a landslide, involves a large amount of ice, snow and rock falling quickly down the side of a mountain.
- A pyroclastic flow is caused by a collapsing cloud of hot ash, gas and rocks from a volcanic explosion that moves rapidly down an erupting volcano.

LANDSLIDE PREDICTION MAPPING

Landslide hazard analysis and mapping can provide useful information for catastrophic loss reduction, and assist in the development of guidelines for sustainable land use planning. The analysis is used to identify the factors that are related to landslides, estimate the relative contribution of factors causing slope failures, establish a relation between the factors and landslides, and to predict the landslide hazard in the future based on such a relationship. The factors that have been used for landslide hazard analysis can usually be grouped intogeomorphology, geology, land use/land cover, and hydrogeology. Since many factors are considered for landslide hazard mapping, GIS is an appropriate tool because it has functions of collection, storage, manipulation, display, and analysis of large amounts of spatially referenced data which can be handled fast and effectively. Remote sensing techniques are also highly employed for landslide hazard assessment and analysis. Before and after aerial photographs and satellite imagery are used to gather landslide characteristics, like distribution and classification, and factors like slope, lithology, and land use/land cover to be used to help predict future events. Before and after imagery also helps to reveal how the landscape changed after an event, what may have triggered the landslide, and shows the process of regeneration and recovery.

Using satellite imagery in combination with GIS and on-the-ground studies, it is possible to generate maps of likely occurrences of future landslides. Such maps should show the locations of previous events as well as clearly indicate the probable locations of future events. In general, to predict landslides, one must assume that their occurrence is determined by certain geologic factors, and that future landslides will occur under the same conditions as past events. Therefore, it is necessary to establish a relationship between the geomorphologic conditions in which the past events took place and the expected future conditions.

Natural disasters are a dramatic example of people living in conflict with the environment. Early predictions and warnings are essential for the reduction of property damage and loss of life. Because landslides occur frequently and can represent some of the most destructive forces on earth, it is imperative to have a good understanding as to what causes them and how people can either help prevent them from occurring or simply avoid them when they do occur. Sustainable land management and development is an essential key to reducing the negative impacts felt by landslides.

GIS offers a superior method for landslide analysis because it allows one to capture, store, manipulate, analyze, and display large amounts of data quickly and effectively. Because so many variables are involved, it is important to be able to overlay the many layers of data to develop a full and accurate portrayal of what is taking place on the Earth's surface. Researchers need to know which variables are the most important factors that trigger landslides in any given location. Using GIS, extremely detailed maps can be generated to show past events and likely future events which have the potential to save lives, property, and money.

Prehistoric landslides

- ⊙ Landslide which moved Heart Mountain to its current location, the largest ever discovered on land. In the 48 million years since the slide occurred, erosion has removed most of the portion of the slide.

- Flims Rockslide, ca. 12 km3 (2. 9 cu mi), Switzerland, some 10000 years ago in post-glacialPleistocene/Holocene, the largest so far described in the alps and on dry land that can be easily identified in a modestly eroded state.

- The landslide around 200BC which formed Lake Waikaremoana on the North Island of New Zealand, where a large block of the Ngamoko Range slid and dammed a gorge of Waikaretaheke River, forming a natural reservoir up to 256 metres deep.

- Cheekye Fan, British Columbia, Canada, ca. 25 km2 (9. 7 sq mi), Late Pleistocene in age.

Extraterrestrial landslides

Evidence of past landslides has been detected on many bodies in the solar system, but since most observations are made by probes that only observe for a limited time and most bodies in the solar system appear to be geologically inactive not many landslides are known to have happened in recent times. Both Venus and Mars have been subject to long-term mapping by orbiting satellites, and examples of landslides have been observed on both.

GLACIAL DEPOSITS 17

An advancing ice sheet carries an abundance of rock that was plucked from the underlying bedrock; only a small amount is carried on the surface from mass wasting. The rock/sediment load of alpine glaciers, on the other hand, comes mostly from rocks that have fallen onto the glacier from the valley walls. The various unsorted rock debris and sediment that is carried or later deposited by a glacier is called till. Till particles typically range from clay sized to boulder sized but can sometimes weigh up to thousands of tons. Boulders that have been carried a considerable distance and then deposited by a glacier are called erratics. Erratics can be a key to determining the direction of movement if the original source of the boulder can be located.

FEATURES LEFT BY VALLEY GLACIERS AND ICE SHEETS

Moraines left by valley glaciers are shown in Figure 1, and features left by a receding ice sheet are shown in Figure 2. Moraines are deposits of till that are left behind when a glacier recedes or that are carried on top of alpine glaciers.

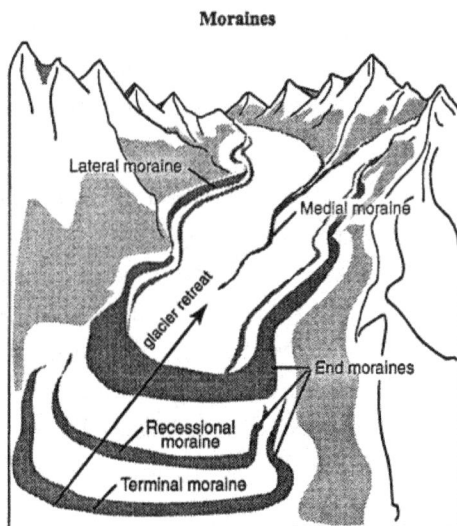

Moraines

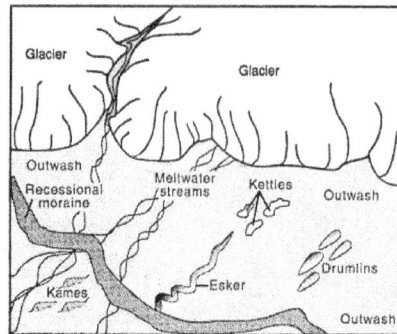

Landforms Produced After Ice Recession

Lateral moraines consist of rock debris and sediment that have worked loose from the walls beside a valley glacier and have built up in ridges along the sides of the glacier. Medial moraines are long ridges of till that result when lateral moraines join as two tributary glaciers merge to form a single glacier. As more tributary glaciers join the main body of ice, a series of roughly parallel medial moraines develop on the surface of main glacier.

LANDFORMS PRODUCED AFTER ICE RECESSION

An extensive pile of till called an end moraine can build up at the front of the glacier and is typically crescent shaped. Two kinds of end moraines are recognized: terminal and recessional moraines. A terminal moraine is the ridge of till that marks the farthest advance of the glacier before it started to recede. A recessional moraine is one that develops at the front of the receding glacier; a series of recessional moraines mark the path of a retreating glacier.

A thin, widespread layer of till deposited across the surface as an ice sheet melts is called a ground moraine. Ground moraine material can sometimes be reshaped by subsequent glaciers into streamlined hills called drumlins, long, narrow, rounded ridges of till whose long axes parallel the direction the glacier traveled.

As a glacier melts, till is released from the ice into the flowing water. The sediments deposited by glacial meltwater are called outwash. Since they have been transported by running water, the outwash deposits are braided, sorted, and layered. The broad front of outwash associated with an ice sheet is called anoutwash plain; if it is from an alpine glacier it is called a valley train. Kames are steep sided mounds of stratified till that were deposited by meltwater in depressions or openings in the ice or as short lived deltas or fans at the mouths of meltwater streams. The rapid build up of sediments can bury isolated blocks of ice. When the ice melts, the resultant depression is called a kettle. Kettle lakes, common in the upper Midwest of the United States, are bodies of water that occupy kettles.

Eskers are long, winding ridges of outwash that were deposited in streams flowing through ice caves and tunnels at the base of the glacier. Generally well sorted and cross bedded, esker sands and gravels eventually choke off the waterway.

The great volume of meltwater often results in the formation of glacial lakes between the end moraines and the retreating glacier front. The sediments that form at the bottom of the lake consist of fine grained silt and clay that have an alternating light dark layering. A varve consists of one light colored bed and one dark colored bed that represent a single year's deposition. The light colored layer is mostly silt that was deposited rapidly during the summer months; the dark layer consists of clay and organic material that formed during the winter. The age of a glacial lake can be determined from the number of varves that have formed on the lake bottom.

Glacier

A glacier (US /ˈɡleɪʃɚ/ or UK /ˈɡlæsiɚ/) is a persistent body of dense ice that is constantly moving under its own weight; it forms where the accumulation of snow exceeds its ablation (melting andsublimation) over many years, often centuries. Glaciers slowly deform and flow due to stresses induced by their weight, creating crevasses, seracs, and other distinguishing features. They also abrade rock and debris from their substrate to create landforms such as cirques and moraines. Glaciers form only on land and are distinct from the much thinner sea ice and lake ice that form on the surface of bodies of water.

On Earth, 99% of glacial ice is contained within vast ice sheets in the polar regions, but glaciers may be found in mountain ranges on every continent except Australia, and on a few high-latitude oceanic islands. Between 35°N and 35°S, glaciers occur only in the Himalayas, Andes, Rocky Mountains, a few high mountains in East Africa, Mexico, New Guinea and on Zard Kuh in Iran.

Glacial ice is the largest reservoir of freshwater on Earth. Many glaciers from temperate, alpine and seasonal polar climates store water as ice during the colder seasons and release it later in the form ofmeltwater as warmer summer temperatures cause the glacier to melt, creating a water source that is especially important for plants, animals and human uses when other sources may be scant. Within high altitude and Antarctic environments, the seasonal temperature difference is often not sufficient to release meltwater.

Because glacial mass is affected by long-term climate changes, e.g., precipitation, mean temperature, and cloud cover, glacial mass changes are considered among the most sensitive indicators of climate change and are a major source of variations in sea level.

A large piece of compressed ice, or a glacier, would appear blue as large quantities of water appear blue. The latter is because the water molecule absorbs other colors more efficiently than blue. The other reason for the blue color of glaciers is the lack of air bubbles. The air bubbles, which give a white color to the regular ice, are squeezed out by pressure increasing the density of the created ice.

ETYMOLOGY AND RELATED TERMS

The word glacier comes from French. It is derived from the Vulgar Latin glacia and ultimately from Latin glacies meaning ice. The processes and features caused by glaciers and related to them are referred to as glacial. The process of glacier establishment, growth and flow is calledglaciation. The corresponding area of study is called glaciology. Glaciers are important components of the global cryosphere.

Types

Glaciers are categorized by their morphology, thermal characteristics, and behavior. Alpine glaciers, also known as mountain glaciers or cirque glaciers, form on the crests and slopes of mountains. An alpine glacier that fills a valley is sometimes called a valley glacier. A large body of glacial ice astride a mountain, mountain range, or volcano is termed an ice cap or ice field. Ice caps have an area less than 50, 000 km^2 (20, 000 mile^2) by definition.

Glacial bodies larger than 50, 000 km^2 are called ice sheets or continental glaciers. Several kilometers deep, they obscure the underlying topography. Only nunataks protrude from their surfaces. The only extant ice sheets are the two that cover most of Antarctica and Greenland. They contain vast quantities of fresh water, enough that if both melted, global sea levels would rise by over 70 meters. Portions of an ice sheet or cap that extend into water are called ice shelves; they tend to be thin with limited slopes and reduced velocities. Narrow, fast-moving sections of an ice sheet are called ice streams. In Antarctica, many ice streams drain into large ice shelves. Some drain directly into the sea, often with an ice tongue, like Mertz Glacier.

Tidewater glaciers are glaciers that terminate in the sea, including most glaciers flowing from Greenland, Antarctica, Baffin and Ellesmere Islands in Canada, Southeast Alaska, and the Northernand Southern Patagonian Ice Fields. As the ice reaches the sea, pieces break off, or calve, formingicebergs. Most tidewater glaciers calve above sea level, which often results in a tremendous impact as the iceberg strikes the water. Tidewater glaciers undergo centuries-long cycles of advance and retreatthat are much less affected by the climate change than those of other glaciers.

Thermally, a temperate glacier is at melting point throughout the year, from its surface to its base. The ice of a polar glacier is always below freezing point from the surface to its base, although the surface snowpack may experience seasonal melting. A sub-polar glacier includes both temperate and polar ice, depending on depth beneath the surface and position along the length of the glacier. In a similar way, the thermal regime of a glacier is often described by the temperature at its base alone. A cold-based glacier is below freezing at the ice-ground interface, and is thus frozen to the underlying substrate. A warm-based glacier is above or at freezing at the interface, and is able to slide at this contact. This contrast is thought to a large extent to govern the ability of a glacier to effectively erode its bed, as sliding ice promotes plucking at rock from the surface below. Glaciers which are partly cold-based and partly warm-based are known as polythermal.

Formation

Glaciers form where the accumulation of snow and ice exceeds ablation. The area in which a glacier forms is called a cirque (corrie or cwm) - a typically armchair-shaped geological feature (such as a depression between mountains enclosed by arêtes) - which collects and compresses through gravity the snow which falls into it. This snow collects and is compacted by the weight of the snow falling above it forming névé. Further crushing of the individual snowflakes and squeezing the air from the snow turns it into 'glacial ice'. This glacial ice will fill the cirque until it 'overflows' through a geological weakness or vacancy, such as the gap between two mountains. When the mass of snow and ice is sufficiently thick, it begins to move due to a combination of surface slope, gravity and pressure. On steeper slopes, this can occur with as little as 15 m (50 ft) of snow-ice.

In temperate glaciers, snow repeatedly freezes and thaws, changing into granular ice called firn. Under the pressure of the layers of ice and snow above it, this granular ice fuses into denser and denser firn. Over a period of years, layers of firn undergo further compaction and become glacial ice. Glacier ice is slightly less dense than ice formed from frozen water because it contains tiny trapped air bubbles.

Glacial ice has a distinctive blue tint because it absorbs some red light due to an overtone of the infrared OH stretching mode of the water molecule. Liquid water is blue for the same reason. The blue of glacier ice is sometimes misattributed to Rayleigh scattering due to bubbles in the ice.

Structure

A glacier originates at a location called its glacier head and terminates at its glacier foot, snout, orterminus.

Glaciers are broken into zones based on surface snowpack and melt conditions. The ablation zone is the region where there is a net loss in glacier mass. The equilibrium line separates the ablation zone and the accumulation zone; it is the altitude where the amount of new snow gained by accumulation is equal to the amount of ice lost through ablation. The upper part of a glacier, where accumulation exceeds ablation, is called the accumulation zone. In general, the accumulation zone accounts for 60–70% of the glacier's surface area, more if the glacier calves icebergs. Ice in the accumulation zone is deep enough to exert a downward force that erodes underlying rock. After a glacier melts, it often leaves behind a bowl- or amphitheater-shaped depression that ranges in size from large basins like the Great Lakes to smaller mountain depressions known ascirques.

The accumulation zone can be subdivided based on its melt conditions.

- The dry snow zone is a region where no melt occurs, even in the summer, and the snowpack remains dry.

- The percolation zone is an area with some surface melt, causing meltwater to percolate into the snowpack. This zone is often marked by refrozen ice lenses, glands, and layers. The snowpack also never reaches melting point.

- Near the equilibrium line on some glaciers, a superimposed ice zone develops. This zone is where meltwater refreezes as a cold layer in the glacier, forming a continuous mass of ice.

- The wet snow zone is the region where all of the snow deposited since the end of the previous summer has been raised to 0 °C.

The health of a glacier is usually assessed by determining the glacier mass balance or observing terminus behavior. Healthy glaciers have large accumulation zones, more than 60% of their area snowcovered at the end of the melt season, and a terminus with vigorous flow.

Following the Little Ice Age's end around 1850, glaciers around the Earth have retreated substantially. A slight cooling led to the advance of many alpine glaciers between 1950–1985, but since 1985 glacier retreat and mass loss has become larger and increasingly ubiquitous.

Motions

Glaciers move, or flow, downhill due to gravity and the internal deformation of ice. Ice behaves like a brittle solid until its thickness exceeds about 50 m (160 ft). The pressure on ice deeper than 50 m causes plastic flow. At the molecular level, ice consists of stacked layers of molecules with relatively weak bonds between layers. When the stress on the layer above exceeds the inter-layer binding strength, it moves faster than the layer below.

Glaciers also move through basal sliding. In this process, a glacier slides over the terrain on which it sits, lubricated by the presence of liquid water. The water is created from ice that melts under high pressure from frictional heating. Basal sliding is dominant in temperate, or warm-based glaciers.

Fracture Zone and Cracks

The top 50 metres (160 ft) of a glacier are rigid because they are under low pressure. This upper section is known as the fracture zone; it mostly moves as a single unit over the plastically flowing lower section. When a glacier moves through irregular terrain, cracks called crevasses develop in the fracture zone. Crevasses form due to differences in glacier velocity. If two rigid sections of a glacier move at different speeds and directions, shear forces cause them to break apart, opening a crevasse. Crevasses are seldom more than 150 feet (46 m) deep but in some cases can be 1, 000 feet (300 m) or even deeper. Beneath this point, the plasticity of the ice is too great for cracks to form. Intersecting crevasses can create isolated peaks in the ice, called seracs.

Crevasses can form in several different ways. Transverse crevasses are transverse to flow and form where steeper slopes cause a glacier to accelerate. Longitudinal crevasses form semi-parallel to flow where a glacier expands laterally. Marginal crevasses form from the edge of the glacier, due to the reduction in speed caused by friction of the valley walls. Marginal crevasses are usually largely transverse to flow. Moving glacier ice can sometimes separate from stagnant ice above, forming abergschrund. Bergschrunds resemble crevasses but are singular features at a glacier's margins.

Crevasses make travel over glaciers hazardous, especially when they are hidden by fragile snow bridges.

Below the equilibrium line, glacial meltwater is concentrated in stream channels. Meltwater can pool in proglacial lakes on top of a glacier or descend into the depths of a glacier via moulins. Streams within or beneath a glacier flow in englacial or sub-glacial tunnels. These tunnels sometimes reemerge at the glacier's surface.

Speed

The speed of glacial displacement is partly determined by friction. Friction makes the ice at the bottom of the glacier move more slowly than ice at the top. In alpine glaciers, friction is also generated at the valley's side walls, which slows the edges relative to the center.

Mean speeds vary greatly, but is typically around 1 meter per day. There may be no motion in stagnant areas; for example, in parts of Alaska, trees can establish themselves on surface sediment deposits. In other cases, glaciers can move as fast as 20–30 m per day, such as in Greenland's Jakobshavn Isbræ (Greenlandic:Sermeq Kujalleq). Velocity increases with increasing slope, increasing thickness, increasing snowfall, increasing longitudinal confinement, increasing basal temperature, increasing meltwater production and reduced bed hardness.

A few glaciers have periods of very rapid advancement called surges. These glaciers exhibit normal movement until suddenly they accelerate, then return to their previous state. During these surges, the glacier may reach velocities far greater than normal speed. These surges may be caused by failure of the underlying bedrock, the pooling of meltwater at the base of the glacier — perhaps delivered from a supraglacial lake — or the simple accumulation of mass beyond a critical "tipping point".

In glaciated areas where the glacier moves faster than one km per year, glacial earthquakes occur. These are large scale temblors that have seismic magnitudes as high as 6. 1. The number of glacial earthquakes in Greenland peaks every year in July, August and September and is increasing over time. In a study using data from January 1993 through October 2005, more events were detected every year since 2002, and twice as many events were recorded in 2005 as there were in any other year. This increase in the numbers of glacial earthquakes in Greenland may be a response to global warming.

Ogives

Ogives are alternating wave crests and valleys that appear as dark and light bands of ice on glacier surfaces. They are linked to seasonal motion of glaciers; the width of one dark and one light band generally equals the annual movement of the glacier. Ogives are formed when ice from an icefall is severely broken up, increasing ablation surface area during summer. This creates a swale and space for snow accumulation in the winter, which in turn creates a ridge. Sometimes ogives consist only of undulations or color bands and are described as wave ogives or band ogives.

Geography

Glaciers are present on every continent and approximately fifty countries, excluding those (Australia, South Africa) that have glaciers only on distant subantarctic island territories. Extensive glaciers are found in Antarctica, Chile, Canada, Alaska, Greenland and Iceland. Mountain glaciers are widespread, especially in the Andes, the Himalayas, the Rocky Mountains, the Caucasus, and the Alps. Mainland Australia currently contains no glaciers, although a small glacier on Mount Kosciuszko was present in the last glacial period. In New Guinea, small, rapidly diminishing, glaciers are located on its highest summit massif of Puncak Jaya. Africa has glaciers on Mount Kilimanjaro in Tanzania, on Mount Kenya and in the Rwenzori Mountains. Oceanic islands with glaciers occur on Iceland, Svalbard, New Zealand, Jan Mayen and the subantarctic islands of Marion, Heard, Grande Terre (Kerguelen) andBouvet. During glacial periods of the Quaternary, Taiwan, Hawaii on Mauna Kea and Tenerife also had large alpine glaciers, while the Faroe and Crozet Islands were completely glaciated.

The permanent snow cover necessary for glacier formation is affected by factors such as the degree of slope on the land, amount of snowfall and the winds. Glaciers can be found in all latitudes except from 20° to 27° north and south of the equator where the presence of the descending limb of the Hadley circulation lowers precipitation so much that with high insolation snow lines reach above 6, 500 metres (21, 330 ft).

Between 19ÚN and 19ÚS, however, precipitation is higher and the mountains above 5, 000 metres (16, 400 ft) usually have permanent snow.

Even at high latitudes, glacier formation is not inevitable. Areas of the Arctic, such as Banks Island, and the McMurdo Dry Valleys in Antarctica are considered polar deserts where glaciers cannot form because they receive little snowfall despite the bitter cold. Cold air, unlike warm air, is unable to transport much water vapor. Even during glacial periods of the Quaternary, Manchuria, lowland Siberia, and central andnorthern Alaska, though extraordinarily cold, had such light snowfall that glaciers could not form.

In addition to the dry, unglaciated polar regions, some mountains and volcanoes in Bolivia, Chile and Argentina are high (4, 500 metres (14, 800 ft) - 6, 900 m (22, 600 ft)) and cold, but the relative lack of precipitation prevents snow from accumulating into glaciers. This is because these peaks are located near or in the hyperarid Atacama Desert.

Glacial Geology

Glaciers erode terrain through two principal processes: abrasion and plucking.

As glaciers flow over bedrock, they soften and lift blocks of rock into the ice. This process, called plucking, is caused by subglacial water that penetrates fractures in the bedrock and subsequently freezes and expands. This expansion causes the ice to act as a lever that loosens the rock by lifting it. Thus, sediments of all sizes become part of the glacier's load. If a retreating glacier gains enough debris, it may become a rock glacier, like the Timpanogos Glacier in Utah.

Abrasion occurs when the ice and its load of rock fragments slide over bedrock and function as sandpaper, smoothing and polishing the bedrock below. The pulverized rock this process produces is called rock flour and is made up of rock grains between 0. 002 and 0. 00625 mm in size. Abrasion leads to steeper valley walls and mountain slopes in alpine settings, which can cause avalanches and rock slides. These add even more material to the glacier.

Glacial abrasion is commonly characterized by glacial striations. Glaciers produce these when they contain large boulders that carve long scratches in the bedrock. By mapping the direction of the striations, researchers can determine the direction of the glacier's movement.

Similar to striations arechatter marks, lines of crescent-shape depressions in the rock underlying a glacier. They are formed by abrasion when boulders in the glacier are repeatedly caught and released as they are dragged along the bedrock.

The rate of glacier erosion is variable. Six factors control erosion rate:

- ◉ Velocity of glacial movement

- ◉ Thickness of the ice

- ◉ Shape, abundance and hardness of rock fragments contained in the ice at the bottom of the glacier

- ◉ Relative ease of erosion of the surface under the glacier

- ◉ Thermal conditions at the glacier base

- ◉ Permeability and water pressure at the glacier base

Material that becomes incorporated in a glacier is typically carried as far as the zone of ablation before being deposited. Glacial deposits are of two distinct types:

- ◉ Glacial till: material directly deposited from glacial ice. Till includes a mixture of undifferentiated material ranging from clay size to boulders, the usual composition of a moraine.

- ◉ Fluvial and outwash sediments: sediments deposited by water. These deposits are stratified by size.

Larger pieces of rock that are encrusted in till or deposited on the surface are called "glacial erratics". They range in size from pebbles to boulders, but as they are often moved great distances, they may be drastically different from the material upon which they are found. Patterns of glacial erratics hint at past glacial motions.

MORAINES

Glacial moraines are formed by the deposition of material from a glacier and are exposed after the glacier has retreated. They usually appear as linear mounds of till, a non-sorted mixture of rock, gravel and boulders within a matrix of a fine powdery material. Terminal or end moraines are formed at the foot or terminal end of a glacier. Lateral moraines are formed on the sides of the glacier. Medial moraines are formed when two different glaciers merge and the lateral moraines of each coalesce to form a moraine in the middle of the combined glacier. Less apparent are ground moraines, also calledglacial drift, which often blankets the surface underneath the glacier downslope from the equilibrium line.

The term moraine is of French origin. It was coined by peasants to describe alluvial embankments and rims found near the margins of glaciers in the French Alps. In modern geology, the term is used more broadly, and is applied to a series of formations, all of which are composed of till. Moraines can also create moraine dammed lakes.

Drumlins

Drumlins are asymmetrical, canoe shaped hills made mainly of till. Their heights vary from 15 to 50 meters and they can reach a kilometer in length. The steepest side of the hill faces the direction from which the ice advanced (stoss), while a longer slope is left in the ice's direction of movement (lee).

Drumlins are found in groups called drumlin fields or drumlin camps. One of these fields is found east of Rochester, New York; it is estimated to contain about 10, 000 drumlins.

Although the process that forms drumlins is not fully understood, their shape implies that they are products of the plastic deformation zone of ancient glaciers. It is believed that many drumlins were formed when glaciers advanced over and altered the deposits of earlier glaciers.

Glacial valleys, cirques, arêtes, and pyramidal peaks

Before glaciation, mountain valleys have a characteristic "V" shape, produced by eroding water. During glaciation, these valleys are widened, deepened, and smoothed, forming a "U"-shaped glacial valley. The erosion that creates glacial valleys eliminates the spurs of earth that extend across mountain valleys, creating triangular cliffs called truncated spurs. Within glacial valleys, depressions created by plucking and abrasion can be filled by lakes, called paternoster lakes. If a glacial valley runs into a large body of water, it forms a fjord.

Many glaciers deepen their valleys more than their smaller tributaries. Therefore, when glaciers recede, the valleys of the tributary glaciers remain above the main glacier's depression and are called hanging valleys.

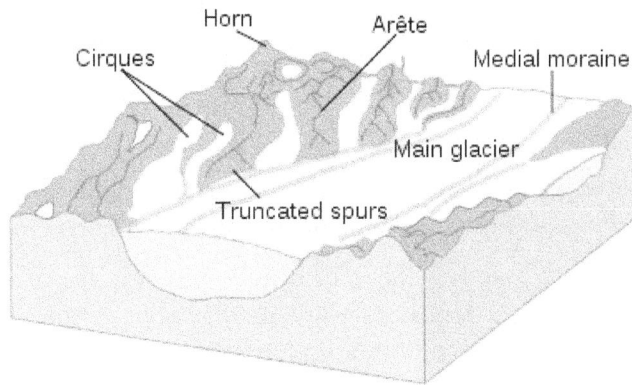

Features of a glacial landscape

At the start of a classic valley glacier is a bowl-shaped cirque, which has escarped walls on three sides but is open on the side that descends into the valley. Cirques are where ice begins to accumulate in a glacier. Two glacial cirques may form back to back and erode their backwalls until only a narrow ridge, called an arête is left. This structure may result in a mountain pass. If multiple cirques encircle a single mountain, they create pointed pyramidal peaks; particularly steep examples are called horns.

Roche moutonnée

Some rock formations in the path of a glacier are sculpted into small hills called roche moutonnée, or "sheepback" rock. Roche moutonnée are elongated, rounded, and asymmetrical bedrock knobs can be produced by glacier erosion. They range in length from less than a meter to several hundred meters long. Roche moutonnée have a gentle slope on their up-glacier sides and a steep to vertical face on their down-glacier sides. The glacier abrades the smooth slope on the upstream side as it flows along, but tears loose and carries away rock from the downstream side via plucking.

Alluvial stratification

As the water that rises from the ablation zone moves away from the glacier, it carries fine eroded sediments with it. As the speed of the water decreases, so does its capacity to carry objects in suspension. The water thus gradually deposits the sediment as it runs, creating an alluvial plain. When this phenomenon occurs in a valley, it is called a valley train. When the deposition is in an estuary, the sediments are known asbay mud.

Outwash plains and valley trains are usually accompanied by basins known as "kettles". These are small lakes formed when large ice blocks that are trapped in alluvium melt and produce water-filled depressions. Kettle diameters range from 5 m to 13 km, with depths of up to 45 meters. Most are circular in shape because the blocks of ice that formed them were rounded as they melted.

Glacial deposits

When a glacier's size shrinks below a critical point, its flow stops and it becomes stationary. Meanwhile, meltwater within and beneath the ice leaves stratified alluvial deposits. These deposits, in the forms of columns, terraces and clusters, remain after the glacier melts and are known as "glacial deposits".

Glacial deposits that take the shape of hills or mounds are called kames. Some kames form when meltwater deposits sediments through openings in the interior of the ice. Others are produced by fans or deltas created by meltwater. When the glacial ice occupies a valley, it can form terraces or kames along the sides of the valley.

Long, sinuous glacial deposits are called eskers. Eskers are composed of sand and gravel that was deposited by meltwater streams that flowed through ice tunnels within or beneath a glacier. They remain after the ice melts, with heights exceeding 100 meters and lengths of as long as 100 km.

Loess deposits

Very fine glacial sediments or rock flour is often picked up by wind blowing over the bare surface and may be deposited great distances from the original fluvial deposition site. These eolian loess deposits may be very deep, even hundreds of meters, as in areas of China and the Midwestern United States of America. Katabatic winds can be important in this process.

Isostatic rebound

Large masses, such as ice sheets or glaciers, can depress the crust of the Earth into the mantle. The depression usually totals a third of the ice sheet or glacier's thickness. After the ice sheet or glacier melts, the mantle begins to flow back to its original position, pushing the crust back up. This post-glacial rebound, which proceeds very slowly after the melting of the ice sheet or glacier, is currently occurring in measurable amounts in Scandinavia and the Great Lakesregion of North America.

A geomorphological feature created by the same process on a smaller scale is known as dilation-faulting. It occurs where previously compressed rock is allowed to return to its original shape more rapidly than can be maintained without faulting. This leads to an effect similar to what would be seen if the rock were hit by a large hammer. Dilation faulting can be observed in recently de-glaciated parts of Iceland and Cumbria.

On Mars

The polar ice caps of Mars show geologic evidence of glacial deposits. The south polar cap is especially comparable to glaciers on Earth. Topographical features and computer models indicate the existence of more glaciers in Mars' past.

At mid-latitudes, between 35° and 65° north or south, Martian glaciers are affected by the thin Martian atmosphere. Because of the low atmospheric pressure, ablation near the surface is solely due tosublimation, not melting. As on Earth, many glaciers are covered with a layer of rocks which insulates the ice. A radar instrument on board the Mars Reconnaissance Orbiter found ice under a thin layer of rocks in formations called lobate debris aprons (LDAs).

COASTAL PROCESSES 18

The shoreline is affected by waves (produced by wind at sea) and tides (produced by the gravitational effect of the moon and sun).

WAVES

Waves are caused by wind. Wave height in the open ocean is determined by three factors. The greater the wind speed the larger the waves. The greater the duration of the wind (or storm) the larger the waves. The greater the fetch (area over which the wind is blowing - size of storm) the larger the waves.

Waves have crests (high spots) and troughs (low spots). The wave height (amplitude) is the difference in height between the crest and the trough. The wavelength (L) is the distance between two crests (or troughs). The period (T) is the time between passage of successive wave crests (or troughs).

Open Ocean Waves: As a wave passes, water molecules rise up and move forward (in the direction of wave motion) until the crest passes. After the crest the water molecules move down and backward. The result is that water molecules move in orbital paths as waves pass. This orbital motion is greatest at the sea surface and decreases with depth below the surface. At a depth of one-half the wavelength the orbital waver motion is nearly zero (actually 4% of the surface orbital diameter). This L/2 depth is considered wave base.

Waves at the Shoreline: As a wave approaches the shore it slows down from drag on the bottom when water depth is less than half the wavelength (L/2). The waves get closer together and taller. Orbital motions of water molecules becomes increasingly elliptical, especially on the bottom. There is a growing proportion of back and forth motion and less up and down motion as the wave moves through shallower and shallower water. Eventually the bottom of the wave slows drastically and the wave topples over as a breaker. The dominant sense of motion is now forward and backwards resulting in the forward swash of water followed by the backwash.

Coastal Sedimentation: In the wave dominated shoreface and nearshore environment fine sediments (silt and clay) remain suspended and are winnowed away, leaving behind the coarse grain sediments (sand and gravel). Suspended fine grain sediments are deposited offshore in deeper water where the bottom is stirred only slightly as depth approaches $L/2$ or not at all by waves (below $L/2$ depth). Therefore sand and gravel dominates the beach, foreshore, near shore. Sands become finer the deeper the water the farther offshore. Sandy bottoms give way to silt-dominated and then clay-dominated bottom sediments (muds) farther offshore and out across the continental shelf.

As a wave crashes on the shore, the water pushes sediment up the beach and then pulls it back down the beach as the water slides back down. If the waves do not come in parallel to the beach longshore transport (littoral drift) of sand occurs. When waves approach the beach at an angle, the part of the wave that reaches shallow water earliest slows down the most, allowing the part of the wave that is farther offshore to catch up. In this way the wave is refracted (bent) so that it crashes on the shore more nearly parallel to the shore. You will never see a wave wash up on a beach at a very high angle from the line of the beach except perhaps at an inlet or where the shore makes a sudden right angle bend. This wave refraction focuses wave energy around headlands and diffuses it in bays. Relatively large waves are found around headlands. Bays have quiet water (good for ship moorings) and are sites of deposition (nice sandy beaches).

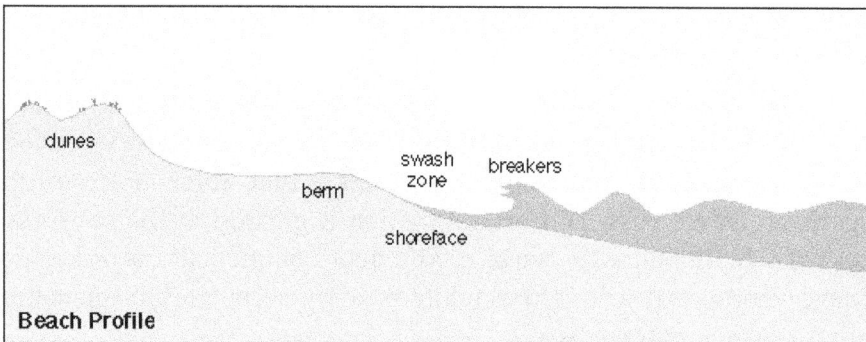

Beach Profile: Summer/Winter Profiles

During calm summer weather with waves gently lapping the shore the beach grows in size. Waves surge up the shoreface. The swash carries sediment. The swash slows, runs out of momentum, then slides back down toward the water. Some of the backwash sinks into the sand. The backwash has a little less energy to carry sediment down the beach so the beach gradually grows in size with the development of a summer berm. The summer beach profile is broader and with a more gentle slope.

During the stormy winter months, storm waves carry much energy to the beach with extra energy to suspend sediments and redistribute them in the nearshore environment. Steady strong winds from a storm can push water up on the leeward shore raising water levels. Return flow from this wind setuphelps to carry sediment away from the shore. The summer berm is eroded away and the sands deposited offshore. The winter beach profile is steeper and narrower.

Tides

Tides result from the gravitational attraction of the sun and the Moon on the oceans. Tidal force is proportional to the mass of the attracting body and inversely proportional to the cube of the distance (tidal force μ mass / distance3). So even though the sun is much much more massive than the moon it is also much farther away. The net result is that the moon exerts about twice the tide force that the sun does.

Equilibrium Theory of Tides:

The Earth and Moon revolve around the center of mass of the Earth-Moon system which is called the barycenter. The gravitational attraction of the Moon pulls up a bulge of water on the side of the moon facing the Earth. The centrifugal force of the Earth's revolution about the barycenter causes a second bulge of water on the side of the Earth opposite the Moon (the Moon's gravitational forces are weaker on the far side of the Earth and the inertial or centrifugal effect is stronger). So as the Earth rotates on its axis, for any observer on the Earth there should be two high tides and two intervening low tides per day. The Moon rises about an hour later every day because it orbits the Earth in the same direction that the Earth spins on its axis. Therefore, the tides should be about an hour later every day.

As the Moon orbits around the Earth every four weeks, the relationship of the Sun, Earth, and Moon changes. At the full moon and new moon the Sun, Earth, and Moon are aligned. At these times the gravitational and centrifugal force of the Moon and Sun combine together. The resulting spring tides are the highest high tides and lowest low tides or the greatest tidal range during the course of the lunar month. At the first quarter and last quarter, the Sun, Earth, and Moon form a right angle. At these times the gravitational and centrifugal forces of the Sun and Moon act at right angles to one another. The resulting neap tides are the lowest high tides and highest low tides or the smallest tidal range.

Problems with the Equilibrium Theory: The equilibrium theory of tides requires that the Earth be covered with an ocean several thousand miles deep covering the entire Earth. Imagine the tidal bulges and troughs as waves on this ocean. Since the rotational speed at the equator is around 1000 miles per hour (25, 000 miles per 24 hours) these waves would have to travel from east to west at 1000 miles per hour across the face of the Earth, at least near the equator. Also consider that the wavelength of one of these waves must be half of 25, 000 miles. Bulge to bulge is 12, 500 miles. Therefore wavebase (L/2) would be 6250 miles. A world encircling ocean would have to be at least this deep or the wave would be slowed by friction with the bottom and would not keep up with the Earth's rotation. Since the Earth's oceans are only on the order of 3 miles rather than 6000 miles deep and since continents litter the surface, the actual tidal pattern is not nearly as simple as the equilibrium model. Nevertheless, most places do have two high and two low tides per day and spring tides and neap tides at the expected times of the month.

Dynamical Theory of Tides:In reality the global tides are organized into a number of tidal cells. Tidal crests rotate once every 12 hours around amphidromic points, counterclockwise in the northern hemisphere, clockwise in the southern hemisphere.

Distance to Moon and Sun: The orbits of the Earth about the Sun and the Moon about the Earth are not circular. They are elliptical. During its monthly orbit about the Earth, the Moon is sometimes closer and sometimes farther from the Earth. The Moon's tidal attraction is strongest at perigee (when the Moon is closest) and weakest at apogee (when the Moon is farthest). The Earth is closer to the sun during the northern hemisphere winter and farther from the sun during the summer. The Sun's tidal attraction is strongest at perihelion (when the Earth is closest to the Sun) and weakest at aphelion (when the Earth is farthest from the Sun).

COASTAL EROSION

Sea Level Rise and Coastal Erosion

Sea level has risen by about 120 m (400 ft) since the peak of the last ice age about 19, 000 years ago. The coast was about 100 km (60 mi) farther offshore from Long Island at that time. Since then ice caps have shrunk, returning water to the sea and the seas have warmed and expanded. Global average sea level is currently rising at the rate of about 2 mm per year. This slow sea level rise helps to increase the rate of coastal erosion. For example, along Long Island's south shore on Fire Island the shoreline is receding between 0. 7 and 1 meter per year.

Coastal flooding and erosion may be severe when storms strike a coast at spring tide. Storms brought by low pressure systems may give rise to a largestorm surge high water resulting from the combination of wind setup and an inverted barometer effect (low pressure doesn't weigh down as heavily on the surface of the ocean). Highest water occurs through the combination of storm surge at spring tide, especially at or near perihelion and perigee.

Coastal Erosion Control

Groins are structures built out from the shore at a right angle to the beach in an attempt to stop longshore sand transport. They hold the sand on the upcurrent side of the groin but the downcurrent side of groins faces enhanced erosion because sand transport from upcurrent is halted. Also, some sand is diverted offshore as longshore currents flow into deeper water around the groin. Jetties are groins built at an inlet to halt the migration of sand across the mouth of the inlet.

Seawalls are structures built parallel to the beach to protect buildings. When storm waves strike a seawall, the unspent wave energy is reflected back offshore. This is good for the building, but that extra energy carries sand offshore. Result: the beach is gone. The seawall will eventually be undermined and the building washed away if the sand is not replenished. In the meantime there is no beach for recreation.

Beach Nourishment: Many beaches in developed areas are not allowed to migrate shoreward with the rising sea. Owners of homes and commercial properties attempt to protect them with groins and seawalls. Large stretches of the eastern seaboard of the U. S. have become armored shorelines with little or no beach, perhaps just a thin strip of sand lying in front of a seawall at high tide. Recreational value is much diminished and the threat to buildings during storms is high. A common recourse is beach nourishment. Sand is pumped onto the beach from offshore, thereby rejuvenating the shore. The process of moving huge quantities of sand is quite expensive, costing millions of dollars for a modest project. Beach nourishment is often done by the Army Corps of Engineers, to protect lives and property, and using taxpayer dollars. However, this is usually a very temporary solution. Beach nourishment increases the slope of the beach which makes it easier for continued erosion. The sand that is pumped from offshore is typically finer than the original beach sand making it easier to erode. The result is that much of the new sand may be washed out in one or two major storms. These beaches commonly lose all of this pumped sand within five years or so.

Dune Stabilization: Coastal dunes form an important barrier to strong storm waves. In the past they were leveled and built upon. Elsewhere, they were trampled by beach goers, killing the dune grasses that kept the sand from blowing away. In recent years many dune protection programs have been put in place to enhance this natural line of protection against the sea. While this is wise, it does not form a permanent, demarcation between the land and the sea. As sea level rises and the shoreline recedes, the dunes will naturally recede landward also. But if they are artificially maintained by planting vegetation, the beach itself will continue to narrow until the shoreface impinges on the dunes. Eventually the dunes wash away and so too the property behind them.

GEOLOGY AND CIVIL ENGINEERING 19

Geology is anearth science comprising the study of solid Earth, the rocks of which it is composed, and the processes by which they change. Geology can also refer generally to the study of the solid features of any celestial body (such as the geology of the Moon or Mars).

Geology gives insight into the history of the Earth by providing the primary evidence for plate tectonics, the evolutionary history of life, and past climates. Geology is important for mineral andhydrocarbon exploration and exploitation, evaluating water resources, understanding of natural hazards, the remediation of environmental problems, and for providing insights into past climate change. Geology also plays a role in geotechnical engineering and is a major academic discipline.

GEOLOGIC MATERIALS

The majority of geological data comes from research on solid Earth materials. These typically fall into one of two categories: rock and unconsolidated material.

Rock

There are three major types of rock: igneous, sedimentary, and metamorphic. The rock cycle is an important concept in geology which illustrates the relationships between these three types of rock, and magma. When a rock crystallizes from melt (magma and/or lava), it is an igneous rock. This rock can be weathered and eroded, and then redeposited andlithified into a sedimentary rock, or be turned into a metamorphic rock due to heat and pressure that change the mineral content of the rock which gives it a characteristic fabric. The sedimentary rock can then be subsequently turned into a metamorphic rock due to heat and pressure and is then weathered, eroded, deposited, and lithified, ultimately becoming a sedimentary rock. Sedimentary rock may also be re-eroded and redeposited, and metamorphic rock may also undergo additional metamorphism. All three types of rocks may be re-melted; when this happens, a new magma is formed, from which an igneous rock may once again crystallize.

Rock Cycle

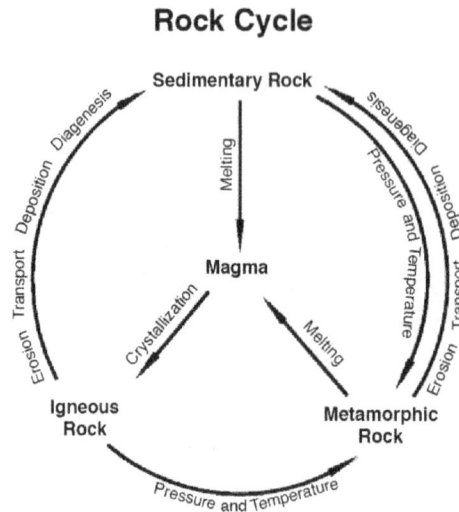

Fig: This schematic diagram of the rock cycle shows the relationship between magma and sedimentary, metamorphic, and igneous rock

The majority of research in geology is associated with the study of rock, as rock provides the primary record of the majority of the geologic history of the Earth.

Unconsolidated Material

Geologists also study unlithified material, which typically comes from more recent deposits. These materials are superficial deposits which lie above the bedrock. Because of this, the study of such material is often known asQuaternary geology, after the recent Quaternary Period. This includes the study of sediment and soils, including studies in geomorphology, sedimentology, and paleoclimatology.

WHOLE-EARTH STRUCTURE

Plate tectonics

In the 1960s, a series of discoveries, the most important of which was seafloor spreading, showed that the Earth's lithosphere, which includes the crust and rigid uppermost portion of the upper mantle, is separated into a number of tectonic plates that move across the plastically deforming, solid, upper mantle, which is called the asthenosphere. There is an intimate coupling between the movement of the plates on the surface and the convection of the mantle: oceanic plate motions and mantleconvection currents always move in the same direction, because the oceanic lithosphere is the rigid upper thermal boundary layer of the convecting mantle. This coupling between rigid plates moving on the surface of the Earth and the convecting mantle is called plate tectonics.

The development of plate tectonics provided a physical basis for many observations of the solid Earth. Long linear regions of geologic features could be explained as plate boundaries. Mid-ocean ridges, high regions on the seafloor where hydrothermal vents and volcanoes exist, were explained as divergent boundaries, where two plates move apart. Arcs of volcanoes and earthquakes were explained asconvergent boundaries, where one plate subducts under another. Transform boundaries, such as the San Andreas fault system, resulted in widespread powerful earthquakes. Plate tectonics also provided a mechanism for Alfred Wegener's theory of continental drift, in which the continents move across the surface of the Earth over geologic time. They also provided a driving force for crustal deformation, and a new setting for the observations of structural geology. The power of the theory of plate tectonics lies in its ability to combine all of these observations into a single theory of how the lithosphere moves over the convecting mantle.

Earth structure

Advances in seismology, computer modeling, andmineralogy and crystallography at high temperatures and pressures give insights into the internal composition and structure of the Earth.

Seismologists can use the arrival times of seismic waves in reverse to image the interior of the Earth. Early advances in this field showed the existence of a liquid outer core (where shear waves were not able to propagate) and a dense solid inner core. These advances led to the development of a layered model of the Earth, with a crust and lithosphere on top, the mantlebelow (separated within itself by seismic discontinuities at 410 and 660 kilometers), and the outer core and inner core below that. More recently, seismologists have been able to create detailed images of wave speeds inside the earth in the same way a doctor images a body in a CT scan. These images have led to a much more detailed view of the interior of the Earth, and have replaced the simplified layered model with a much more dynamic model.

Mineralogists have been able to use the pressure and temperature data from the seismic and modelling studies alongside knowledge of the elemental composition of the Earth to reproduce these conditions in experimental settings and measure changes in crystal structure. These studies explain the chemical changes associated with the major seismic discontinuities in the mantle and show the crystallographic structures expected in the inner core of the Earth.

Geologic time

The geologic time scale encompasses the history of the Earth. It is bracketed at the old end by the dates of the earliest Solar System material at 4. 567 Ga, (gigaannum: billion years ago) and the age of the Earth at 4. 54 Ga at the beginning of the informally recognized Hadean eon. At the young end of the scale, it is bracketed by the present day in the Holocene epoch.

Brief time scale

The following four timelines show the geologic time scale. The first shows the entire time from the formation of the Earth to the present, but this compresses the most recent eon. Therefore the second scale shows the most recent eon with an expanded scale. The second scale compresses the most recent era, so the most recent era is expanded in the third scale. Since the Quaternary is a very short period with short epochs, it is further expanded in the fourth scale. The second, third, and fourth timelines are therefore each subsections of their preceding timeline as indicated by asterisks. The Holocene (the latest epoch) is too small to be shown clearly on the third timeline on the right, another reason for expanding the fourth scale. The Pleistocene (P) epoch. Q stands for the Quaternary period.

Dating Methods

Geologists use a variety of methods to give both relative and absolute dates to geological events. They then use these dates to find the rates at which processes occur.

Relative dating

Methods for relative dating were developed when geology first emerged as a formal science. Geologists still use the following principles today as a means to provide information about geologic history and the timing of geologic events.

The principle of Uniformitarianism states that the geologic processes observed in operation that modify the Earth's crust at present have worked in much the same way over geologic time. A fundamental principle of geology advanced by the 18th century Scottish physician and geologist James Hutton, is that "the present is the key to the past." In Hutton's words: "the past history of our globe must be explained by what can be seen to be happening now. "

The principle of intrusive relationships concerns crosscutting intrusions. In geology, when an igneous intrusion cuts across a formation of sedimentary rock, it can be determined that the igneous intrusion is younger than the sedimentary rock. There are a number of different types of intrusions, including stocks, laccoliths, batholiths, sills anddikes.

The principle of cross-cutting relationships pertains to the formation of faults and the age of the sequences through which they cut. Faults are younger than the rocks they cut; accordingly, if a fault is found that penetrates some formations but not those on top of it, then the formations that were cut are older than the fault, and the ones that are not cut must be younger than the fault. Finding the key bed in these situations may help determine whether the fault is a normal fault or a thrust fault.

The principle of inclusions and components states that, with sedimentary rocks, if inclusions (or clasts) are found in a formation, then the inclusions must be older than the formation that contains them. For example, in sedimentary rocks, it is common for gravel from an older formation to be ripped up and included in a newer layer. A similar situation with igneous rocks occurs when xenoliths are found. These foreign bodies are picked up as magma or lava flows, and are incorporated, later to cool in the matrix. As a result, xenoliths are older than the rock which contains them.

The principle of original horizontality states that the deposition of sediments occurs as essentially horizontal beds. Observation of modern marine and non-marine sediments in a wide variety of environments supports this generalization (although cross-bedding is inclined, the overall orientation of cross-bedded units is horizontal).

The principle of superposition states that a sedimentary rock layer in a tectonically undisturbed sequence is younger than the one beneath it and older than the one above it. Logically a younger layer cannot slip beneath a layer previously deposited. This principle allows sedimentary layers to be viewed as a form of vertical time line, a partial or complete record of the time elapsed from deposition of the lowest layer to deposition of the highest bed.

The principle of faunal succession is based on the appearance of fossils in sedimentary rocks. As organisms exist at the same time period throughout the world, their presence or (sometimes) absence may be used to provide a relative age of the formations in which they are found. Based on principles laid out by William Smith almost a hundred years before the publication of Charles Darwin's theory of evolution, the principles of succession were developed independently of evolutionary thought. The principle becomes quite complex, however, given the uncertainties of fossilization, the localization of fossil types due to lateral changes in habitat (facies change in sedimentary strata), and that not all fossils may be found globally at the same time.

Absolute dating

Geologists also use methods to determine the absolute age of rock samples and geological events. These dates are useful on their own and may also be used in conjunction with relative dating methods or to calibrate relative methods.

At the beginning of the 20th century, important advancement in geological science was facilitated by the ability to obtain accurate absolute dates to geologic events using radioactive isotopes and other methods. This changed the understanding of geologic time. Previously, geologists could only use fossils and stratigraphic correlation to date sections of rock relative to one another. With isotopic dates it became possible to assign absolute ages to rock units, and these absolute dates could be applied to fossil sequences in which there was datable material, converting the old relative ages into new absolute ages.

For many geologic applications, isotope ratios of radioactive elements are measured in minerals that give the amount of time that has passed since a rock passed through its particular closure temperature, the point at which different radiometric isotopes stop diffusing into and out of the crystal lattice. These are used in geochronologic and thermochronologic studies. Common methods include uranium-lead dating, potassium-argon dating, argon-argon dating and uranium-thorium dating. These methods are used for a variety of applications. Dating of lavaand volcanic ash layers found within a stratigraphic sequence can provide absolute age data for sedimentary rock units which do not contain radioavtive isotopes and calibrate relative dating techniques. These methods can also be used to determine ages of pluton emplacement. Thermochemical techniques can be used to determine temperature profiles within the crust, the uplift of mountain ranges, and paleotopography.

Fractionation of the lanthanide series elements is used to compute ages since rocks were removed from the mantle.

Other methods are used for more recent events. Optically stimulated luminescence and cosmogenic radionucleide dating are used to date surfaces and/or erosion rates. Dendrochronology can also be used for the dating of landscapes. Radiocarbon dating is used for geologically young materials containing organic carbon.

Geological Development of an Area

The geology of an area changes through time as rock units are deposited and inserted and deformational processes change their shapes and locations.

Rock units are first emplaced either by deposition onto the surface or intrusion into the overlying rock. Deposition can occur when sediments settle onto the surface of the Earth and later lithify into sedimentary rock, or when as volcanic material such as volcanic ash or lava flows blanket the surface. Igneous intrusions such as batholiths, laccoliths, dikes, and sills, push upwards into the overlying rock, and crystallize as they intrude.

After the initial sequence of rocks has been deposited, the rock units can be deformed and/or metamorphosed. Deformation typically occurs as a result of horizontal shortening, horizontal extension, or side-to-side (strike-slip) motion. These structural regimes broadly relate to convergent boundaries, divergent boundaries, and transform boundaries, respectively, between tectonic plates.

When rock units are placed under horizontal compression, they shorten and become thicker. Because rock units, other than muds, do not significantly change in volume, this is accomplished in two primary ways: through faultingand folding. In the shallow crust, where brittle deformation can occur, thrust faults form, which cause deeper rock to move on top of shallower rock. Because deeper rock is often older, as noted by the principle of superposition, this can result in older rocks moving on top of younger ones. Movement along faults can result in folding, either because the faults are not planar or because rock layers are dragged along, forming drag folds as slip occurs along the fault. Deeper in the Earth, rocks behave plastically, and fold instead of faulting. These folds can either be those where the material in the center of the fold buckles upwards, creating "antiforms", or where it buckles downwards, creating "synforms". If the tops of the rock units within the folds remain pointing upwards, they are called anticlines and synclines, respectively. If some of the units in the fold are facing downward, the structure is called an overturned anticline or syncline, and if all of the rock units are overturned or the correct up-direction is unknown, they are simply called by the most general terms, antiforms and synforms.

Even higher pressures and temperatures during horizontal shortening can cause both folding andmetamorphism of the rocks. This metamorphism causes changes in the mineral composition of the rocks; creates a foliation, or planar surface, that is related to mineral growth under stress. This can remove signs of the original textures of the rocks, such as bedding in sedimentary rocks, flow features of lavas, and crystal patterns in crystalline rocks.

Extension causes the rock units as a whole to become longer and thinner. This is primarily accomplished through normal faulting and through the ductile stretching and thinning. Normal faults drop rock units that are higher below those that are lower. This typically results in younger units being placed below older units. Stretching of units can result in their thinning; in fact, there is a location within the Maria Fold and Thrust Belt in which the entire sedimentary sequence of the Grand Canyon can be seen over a length of less than a meter. Rocks at the depth to be ductilely stretched are often also metamorphosed. These stretched rocks can also pinch into lenses, known as boudins, after the French word for "sausage", because of their visual similarity.

Where rock units slide past one another, strike-slip faults develop in shallow regions, and become shear zones at deeper depths where the rocks deform ductilely.

The addition of new rock units, both depositionally and intrusively, often occurs during deformation. Faulting and other deformational processes result in the creation of topographic gradients, causing material on the rock unit that is increasing in elevation to be eroded by hillslopes and channels. These sediments are deposited on the rock unit that is going down. Continual motion along the fault maintains the topographic gradient in spite of the movement of sediment, and continues to createaccommodation space for the material to deposit. Deformational events are often also associated with volcanism and igneous activity. Volcanic ashes and lavas accumulate on the surface, and igneous intrusions enter from below. Dikes, long, planar igneous intrusions, enter along cracks, and therefore often form in large numbers in areas that are being actively deformed. This can result in the emplacement of dike swarms, such as those that are observable across the Canadian shield, or rings of dikes around the lava tube of a volcano.

All of these processes do not necessarily occur in a single environment, and do not necessarily occur in a single order. The Hawaiian Islands, for example, consist almost entirely of layered basaltic lava flows. The sedimentary sequences of the mid-continental United States and the Grand Canyon in the southwestern United States contain almost-undeformed stacks of sedimentary rocks that have remained in place since Cambrian time. Other areas are much more geologically complex. In the southwestern United States, sedimentary, volcanic, and intrusive rocks have been metamorphosed, faulted, foliated, and folded. Even older rocks, such as the Acasta gneiss of the Slave craton in northwestern Canada, the oldest known rock in the world have been metamorphosed to the point where their origin is undiscernable without laboratory analysis. In addition, these processes can occur in stages. In many places, the Grand Canyon in the southwestern United States being a very visible example, the lower rock units were metamorphosed and deformed, and then deformation ended and the upper, undeformed units were deposited. Although any amount of rock emplacement and rock deformation can occur, and they can occur any number of times, these concepts provide a guide to understanding the geological history of an area.

METHODS OF GEOLOGY

Geologists use a number of field, laboratory, and numerical modeling methods to decipher Earth history and understand the processes that occur on and inside the Earth. In typical geological investigations, geologists use primary information related to petrology (the study of rocks), stratigraphy (the study of sedimentary layers), and structural geology (the study of positions of rock units and their deformation). In many cases, geologists also study modern soils, rivers, landscapes, and glaciers; investigate past and current life and biogeochemical pathways, and use geophysical methods to investigate the subsurface.

Field methods

Geological field work varies depending on the task at hand. Typical fieldwork could consist of:

- ◉ Geological mapping
- ◉ Structural mapping: the locations of major rock units and the faults and folds that led to their placement there.
- ◉ Stratigraphic mapping: the locations of sedimentary facies (lithofacies and biofacies) or the mapping of isopachs of equal thickness of sedimentary rock
- ◉ Surficial mapping: the locations of soils and surficial deposits

- Surveying of topographic features
- Creation of topographic maps
- Work to understand change across landscapes, including:
- Patterns of erosion and deposition
- River channel change through migration and avulsion
- Hillslope processes
- Subsurface mapping through geophysical methods
- These methods include:
- Shallow seismic surveys
- Ground-penetrating radar
- Aeromagnetic surveys
- Electrical resistivity tomography
- They are used for:
- Hydrocarbon exploration
- Finding groundwater
- Locating buried archaeological artifacts
- High-resolution stratigraphy
- Measuring and describing stratigraphic sections on the surface
- Well drilling and logging
- Biogeochemistry and geomicrobiology
- Collecting samples to:
- Determine biochemical pathways
- Identify new species of organisms
- Identify new chemical compounds
- And to use these discoveries to:
- Understand early life on Earth and how it functioned and metabolized
- Find important compounds for use in pharmaceuticals.
- Paleontology: excavation of fossil material
- For research into past life and evolution
- For museums and education
- Collection of samples for geochronology and thermochronology
- Glaciology: measurement of characteristics of glaciers and their motion

Petrology

In addition to identifying rocks in the field, petrologists identify rock samples in the laboratory. Two of the primary methods for identifying rocks in the laboratory are through optical microscopy and by using an electron microprobe. In an optical mineralogy analysis, thin sections of rock samples are analyzed through a petrographic microscope, where the minerals can be identified through their different properties in plane-polarized and cross-polarized light, including their birefringence, pleochroism, twinning, and interference properties with a conoscopic lens. In the electron microprobe, individual locations are analyzed for their exact chemical compositions and variation in composition within individual crystals. Stable and radioactive isotope studies provide insight into thegeochemical evolution of rock units.

Petrologists can also use fluid inclusion data and perform high temperature and pressure physical experiments to understand the temperatures and pressures at which different mineral phases appear, and how they change through igneous and metamorphic processes. This research can be extrapolated to the field to understand metamorphic processes and the conditions of crystallization of igneous rocks. This work can also help to explain processes that occur within the Earth, such assubduction and magma chamber evolution.

STRUCTURAL GEOLOGY

Structural geologists use microscopic analysis of oriented thin sections of geologic samples to observe the fabric within the rocks which gives information about strain within the crystalline structure of the rocks. They also plot and combine measurements of geological structures in order to better understand the orientations of faults and folds in order to reconstruct the history of rock deformation in the area. In addition, they perform analog and numerical experiments of rock deformation in large and small settings.

The analysis of structures is often accomplished by plotting the orientations of various features onto stereonets. A stereonet is a stereographic projection of a sphere onto a plane, in which planes are projected as lines and lines are projected as points. These can be used to find the locations of fold axes, relationships between faults, and relationships between other geologic structures.

Among the most well-known experiments in structural geology are those involving orogenic wedges, which are zones in which mountains are built along convergent tectonic plate boundaries. In the analog versions of these experiments, horizontal layers of sand are pulled along a lower surface into a back stop, which results in realistic-looking patterns of faulting and the growth of a critically tapered (all angles remain the same) orogenic wedge.

Numerical models work in the same way as these analog models, though they are often more sophisticated and can include patterns of erosion and uplift in the mountain belt. This helps to show the relationship between erosion and the shape of the mountain range. These studies can also give useful information about pathways for metamorphism through pressure, temperature, space, and time.

Stratigraphy

In the laboratory, stratigraphers analyze samples of stratigraphic sections that can be returned from the field, such as those from drill cores. Stratigraphers also analyze data from geophysical surveys that show the locations of stratigraphic units in the subsurface. Geophysical data and well logs can be combined to produce a better view of the subsurface, and stratigraphers often use computer programs to do this in three dimensions. Stratigraphers can then use these data to reconstruct ancient processes occurring on the surface of the Earth, interpret past environments, and locate areas for water, coal, and hydrocarbon extraction.

In the laboratory, biostratigraphers analyze rock samples from outcrop and drill cores for the fossils found in them. These fossils help scientists to date the core and to understand the depositional environment in which the rock units formed. Geochronologists precisely date rocks within the stratigraphic section in order to provide better absolute bounds on the timing and rates of deposition. Magnetic stratigraphers look for signs of magnetic reversals in igneous rock units within the drill cores. Other scientists perform stable isotope studies on the rocks to gain information about past climate.

Planetary geology

With the advent of space exploration in the twentieth century, geologists have begun to look at other planetary bodies in the same ways that have been developed to study the Earth. This new field of study is called planetary geology (sometimes known as astrogeology) and relies on known geologic principles to study other bodies of the solar system.

Although the Greek-language-origin prefix geo refers to Earth, "geology" is often used in conjunction with the names of other planetary bodies when describing their composition and internal processes: examples are "the geology of Mars" and "Lunar geology". Specialised terms such as selenology (studies of the Moon), areology (of Mars), etc., are also in use.

Although planetary geologists are interested in studying all aspects of other planets, a significant focus is to search for evidence of past or present life on other worlds. This has led to many missions whose primary or ancillary purpose is to examine planetary bodies for evidence of life. One of these is the Phoenix lander, which analyzed Martian polar soil for water, chemical, and mineralogical constituents related to biological processes.

APPLIED GEOLOGY

Economic geology

Economic geologists help locate and manage the Earth's natural resources, such as petroleum and coal, as well as mineral resources, which include metals such as iron, copper, and uranium.

Mining geology

Mining geology consists of the extractions of mineral resources from the Earth. Some resources of economic interests include gemstones, metals, and many minerals such as asbestos, perlite, mica, phosphates, zeolites, clay, pumice, quartz, and silica, as well as elements such assulfur, chlorine, and helium.

Petroleum geology

Petroleum geologists study the locations of the subsurface of the Earth which can contain extractable hydrocarbons, especially petroleum and natural gas. Because many of these reservoirs are found insedimentary basins, they study the formation of these basins, as well as their sedimentary and tectonic evolution and the present-day positions of the rock units.

Engineering geology

Engineering geology is the application of the geologic principles to engineering practice for the purpose of assuring that the geologic factors affecting the location, design, construction, operation, and maintenance of engineering works are properly addressed.

In the field of civil engineering, geological principles and analyses are used in order to ascertain the mechanical principles of the material on which structures are built. This allows tunnels to be built without collapsing, bridges and skyscrapers to be built with sturdy foundations, and buildings to be built that will not settle in clay and mud.

Hydrology and Environmental Issues

Geology and geologic principles can be applied to various environmental problems such as stream restoration, the restoration of brownfields, and the understanding of the interaction between natural habitat and the geologic environment. Groundwater hydrology, or hydrogeology, is used to locate groundwater, which can often provide a ready supply of uncontaminated water and is especially important in arid regions, and to monitor the spread of contaminants in groundwater wells.

Geologists also obtain data through stratigraphy, boreholes, core samples, and ice cores. Ice cores and sediment cores are used to for paleoclimate reconstructions, which tell geologists about past and present temperature, precipitation, and sea level across the globe. These datasets are our primary source of information on global climate change outside of instrumental data.

Natural hazards

Geologists and geophysicists study natural hazards in order to enact safe building codes and warning systems that are used to prevent loss of property and life. Examples of important natural hazards that are pertinent to geology (as opposed those that are mainly or only pertinent to meteorology) are:

HISTORY OF GEOLOGY

The study of the physical material of the Earth dates back at least to ancient Greece whenTheophrastus (372–287 BCE) wrote the work Peri Lithon (On Stones). During the Roman period, Pliny the Elder wrote in detail of the many minerals and metals then in practical use – even correctly noting the origin of amber.

Some modern scholars, such as Fielding H. Garrison, are of the opinion that modern geology began in the medieval Islamic world. Abu al-Rayhan al-Biruni (973–1048 CE) was one of the earliest Muslim geologists, whose works included the earliest writings on the geology of India, hypothesizing that theIndian subcontinent was once a sea. Islamic Scholar Ibn Sina (Avicenna, 981–1037) proposed detailed explanations for the formation of mountains, the origin of earthquakes, and other topics central to modern geology, which provided an essential foundation for the later development of the science. In China, the polymath Shen Kuo (1031–1095) formulated a hypothesis for the process of land formation: based on his observation of fossil animal shells in a geological stratum in a mountain hundreds of miles from the ocean, he inferred that the land was formed by erosion of the mountains and by deposition of silt.

Nicolas Steno (1638–1686) is credited with the law of superposition, the principle of original horizontality, and the principle of lateral continuity: three defining principles of stratigraphy.

The word geology was first used by Ulisse Aldrovandi in 1603, then by Jean-André Deluc in 1778 and introduced as a fixed term by Horace-Bénédict de Saussure in 1779. The word is derived from theGreek ãÆ, gê, meaning "earth" and ëüãïò, logos, meaning "speech". But according to another source, the word "geology" comes from a Norwegian, Mikkel Pedersøn Escholt (1600–1699), who was a priest and scholar. Escholt first used the definition in his book titled, Geologica Norvegica (1657).

William Smith (1769–1839) drew some of the first geological maps and began the process of ordering rock strata (layers) by examining the fossils contained in them.

James Hutton is often viewed as the first modern geologist. In 1785 he presented a paper entitled Theory of the Earth to the Royal Society of Edinburgh. In his paper, he explained his theory that the Earth must be much older than had previously been supposed in order to allow enough time for mountains to be eroded and for sediments to form new rocks at the bottom of the sea, which in turn were raised up to become dry land. Followers of Hutton were known as Plutonists because they believed that some rocks were formed by vulcanism, which is the deposition of lava from volcanoes, as opposed to the Neptunists, led by Abraham Werner, who believed that all rocks had settled out of a large ocean whose level gradually dropped over time. The first geological map of the U. S. was produced in 1809 by William Maclure. In 1807, Maclure commenced the self-imposed task of making a geological survey of the United States. Almost every state in the Union was traversed and mapped by him, the Allegheny Mountains being crossed and recrossed some 50 times. The results of his unaided labours were submitted to the American Philosophical Society in a memoir entitled Observations on the Geology of the United States explanatory of a Geological Map, and published in the Society's Transactions, together with the nation's first geological map. This antedates William Smith's geological map of England by six years, although it was constructed using a different classification of rocks.

Sir Charles Lyell first published his famous book, Principles of Geology, in 1830. This book, which influenced the thought of Charles Darwin, successfully promoted the doctrine of uniformitarianism. This theory states that slow geological processes have occurred throughout the Earth's history and are still occurring today. In contrast, catastrophism is the theory that Earth's features formed in single, catastrophic events and remained unchanged thereafter. Though Hutton believed in uniformitarianism, the idea was not widely accepted at the time.

Much of 19th-century geology revolved around the question of the Earth's exact age. Estimates varied from a few hundred thousand to billions of years. By the early 20th century, radiometric dating allowed the Earth's age to be estimated at two billion years. The awareness of this vast amount of time opened the door to new theories about the processes that shaped the planet.

Some of the most significant advances in 20th-century geology have been the development of the theory of plate tectonics in the 1960s and the refinement of estimates of the planet's age. Plate tectonics theory arose from two separate geological observations: seafloor spreading and continental drift. The theory revolutionized the Earth sciences. Today the Earth is known to be approximately 4. 5 billion years old.

Civil engineering

Civil engineering is a professional engineering discipline that deals with the design, construction, and maintenance of the physical and naturally built environment, including works like roads, bridges, canals, dams, and buildings. Civil engineering is the second-oldest engineering discipline after military engineering, and it is defined to distinguish non-military engineering from military engineering. It is traditionally broken into several sub-disciplines including architectural engineering, environmental engineering, geotechnical engineering, control engineering, structural engineering, earthquake engineering, transportation engineering, forensic engineering, municipal or urban engineering, water resources engineering, materials engineering, wastewater engineering, offshore engineering, aerospace engineering, facade engineering, quantity surveying, coastal engineering, construction surveying, and construction engineering. Civil engineering takes place in the public sector from municipal through to national governments, and in the private sector from individual homeowners through to international companies.

History of the civil engineering profession

Engineering has been an aspect of life since the beginnings of human existence. The earliest practice of civil engineering may have commenced between 4000 and 2000 BC in Ancient Egypt and Mesopotamia (Ancient Iraq) when humans started to abandon a nomadic existence, creating a need for the construction of shelter. During this time, transportation became increasingly important leading to the development of the wheel andsailing. Until modern times there was no clear distinction between civil engineering and architecture, and the term engineer and architect were mainly geographical variations referring to the same occupation, and often used interchangeably. The construction of pyramids in Egypt (circa 2700–2500 BC) were some of the first instances of large structure constructions. Other ancient historic civil engineering constructions include theQanat water management system (the oldest is older than 3000 years and longer than 71 km,) theParthenon by Iktinos in Ancient Greece (447–438 BC), the Appian Way by Roman engineers (c. 312 BC), theGreat Wall of China by General Meng T'ien under orders from Ch'in Emperor Shih Huang Ti (c. 220 BC) and the stupas constructed in ancient Sri Lanka like the Jetavanaramaya and the extensive irrigation works inAnuradhapura. The Romans developed civil structures throughout their empire, including especially aqueducts, insulae, harbors, bridges, dams and roads.

In the 18th century, the term civil engineering was coined to incorporate all things civilian as opposed to military engineering. The first self-proclaimed civil engineer was John Smeaton, who constructed theEddystone Lighthouse. In 1771 Smeaton and some of his colleagues formed the Smeatonian Society of Civil Engineers, a group of leaders of the profession who met informally over dinner. Though there was evidence of some technical meetings, it was little more than a social society.

In 1818 the Institution of Civil Engineers was founded in London, and in 1820 the eminent engineerThomas Telford became its first president. The institution received a Royal Charter in 1828, formally recognising civil engineering as a profession. Its charter defined civil engineering as:

the art of directing the great sources of power in nature for the use and convenience of man, as the means of production and of traffic in states, both for external and internal trade, as applied in the construction of roads, bridges, aqueducts, canals, river navigation and docks for internal intercourse and exchange, and in the construction of ports, harbours, moles, breakwaters and lighthouses, and in the art of navigation by artificial power for the purposes of commerce, and in the construction and application of machinery, and in the drainage of cities and towns.

The first private college to teach Civil Engineering in the United States was Norwich University, founded in 1819 by Captain Alden Partridge. The first degree in Civil Engineering in the United States was awarded by Rensselaer Polytechnic Institute in 1835. The first such degree to be awarded to a woman was granted by Cornell University to Nora Stanton Blatch in 1905.

History of civil engineering

Civil engineering is the application of physical and scientific principles for solving the problems of society, and its history is intricately linked to advances in understanding of physics andmathematics throughout history. Because civil engineering is a wide ranging profession, including several separate specialized sub-disciplines, its history is linked to knowledge of structures, materials science, geography, geology, soils, hydrology, environment, mechanics and other fields.

Throughout ancient and medieval history most architectural design and construction was carried out by artisans, such as stonemasons andcarpenters, rising to the role of master builder. Knowledge was retained in guilds and seldom supplanted by advances. Structures, roads and infrastructure that existed were repetitive, and increases in scale were incremental.

One of the earliest examples of a scientific approach to physical and mathematical problems applicable to civil engineering is the work ofArchimedes in the 3rd century BC, including Archimedes Principle, which underpins our understanding of buoyancy, and practical solutions such as Archimedes' screw. Brahmagupta, an Indian mathematician, used arithmetic in the 7th century AD, based on Hindu-Arabic numerals, for excavation (volume) computations.

THE CIVIL ENGINEER

Education and licensure

Civil engineers typically possess an academic degree in civil engineering. The length of study is three to five years, and the completed degree is designated as a bachelor of engineering, or abachelor of science. The curriculum generally includes classes in physics, mathematics, project management, design and specific topics in civil engineering. After taking basic courses in most sub-disciplines of civil engineering, they move onto specialize in one or more sub-disciplines at advanced levels. While an undergraduate degree (BEng/BSc) normally provides successful students with industry-accredited qualification, some academic institutions offer post-graduate degrees (MEng/MSc), which allow students to further specialize in their particular area of interest.

In most countries, a bachelor's degree in engineering represents the first step towards professional certification, and a professional body certifies the degree program. After completing a certified degree program, the engineer must satisfy a range of requirements (including work experience and exam requirements) before being certified. Once certified, the engineer is designated as a professional engineer (in the United States, Canada and South Africa), a chartered engineer (in most Commonwealth countries), a chartered professional engineer (in Australia and New Zealand), or a European engineer (in most countries of the European Union). There are international agreements between relevant professional bodies to allow engineers to practice across national borders.

The benefits of certification vary depending upon location. For example, in the United States and Canada, "only a licensed professional engineer may prepare, sign and seal, and submit engineering plans and drawings to a public authority for approval, or seal engineering work for public and private clients. " This requirement is enforced under provincial law such as the Engineers Act in Quebec.

No such legislation has been enacted in other countries including the United Kingdom. In Australia, state licensing of engineers is limited to the state of Queensland. Almost all certifying bodies maintain a code of ethics which all members must abide by.

Engineers must obey contract law in their contractual relationships with other parties. In cases where an engineer's work fails, he may be subject to the law of tort of negligence, and in extreme cases, criminal charges. An engineer's work must also comply with numerous other rules and regulations such as building codes and environmental law.

Sub-disciplines

In general, civil engineering is concerned with the overall interface of human created fixed projects with the greater world. General civil engineers work closely with surveyors and specialized civil engineers to design grading, drainage, pavement, water supply, sewer service, electric and communications supply, and land divisions. General engineers spend much time visiting project sites, developing community consensus, and preparing construction plans. General civil engineering is also referred to as site engineering, a branch of civil engineering that primarily focuses on converting a tract of land from one usage to another. Civil engineers apply the principles of geotechnical engineering, structural engineering, environmental engineering, transportation engineering and construction engineering to residential, commercial, industrial and public works projects of all sizes and levels of construction.

Materials science and engineering

Materials science is closely related to civil engineering. It studies fundamental characteristics of materials, and deals with ceramics such as concrete and mix asphalt concrete, strong metals such as aluminum and steel, and polymers including polymethylmethacrylate (PMMA) and carbon fibers.

Materials engineering involves protection and prevention (paints and finishes). Alloying combines two types of metals to produce another metal with desired properties. It incorporates elements of applied physics and chemistry. With recent media attention on nanoscience andnanotechnology, materials engineering has been at the forefront of academic research. It is also an important part of forensic engineering andfailure analysis.

Coastal engineering

Coastal engineering is concerned with managing coastal areas. In some jurisdictions, the terms sea defense and coastal protection mean defense against flooding and erosion, respectively. The term coastal defense is the more traditional term, but coastal management has become more popular as the field has expanded to techniques that allow erosion to claim land.

Construction engineering

Construction engineering involves planning and execution, transportation of materials, site development based on hydraulic, environmental, structural and geotechnical engineering. As construction firms tend to have higher business risk than other types of civil engineering firms do, construction engineers often engage in more business-like transactions, for example, drafting and reviewing contracts, evaluating logistical operations, and monitoring prices of supplies.

Earthquake engineering

Earthquake engineering involves designing structures to withstand hazardous earthquake exposures. Earthquake engineering is a sub-discipline of structural engineering. The main objectives of earthquake engineering are to understand interaction of structures on the shaky ground; foresee the consequences of possible earthquakes; and design, construct and maintain structures to perform at earthquake in compliance with building codes.

Environmental engineering

Environmental engineering is the contemporary term for sanitary engineering, though sanitary engineering traditionally had not included much of the hazardous waste management and environmental remediation work covered by environmental engineering. Public health engineering and environmental health engineering are other terms being used.

Environmental engineering deals with treatment of chemical, biological, or thermal wastes, purification of water and air, and remediation of contaminated sites after waste disposal or accidental contamination. Among the topics covered by environmental engineering are pollutant transport, water purification, waste water treatment, air pollution, solid waste treatment, and hazardous waste management. Environmental engineers administer pollution reduction, green engineering, and industrial ecology. Environmental engineers also compile information on environmental consequences of proposed actions.

Geotechnical engineering

Geotechnical engineering studies rock and soil supporting civil engineering systems. Knowledge from the field of soil science, materials science, mechanics, and hydraulics is applied to safely and economically design foundations, retaining walls, and other structures. Environmental efforts to protectgroundwater and safely maintain landfills have spawned a new area of research called geoenvironmental engineering.

Identification of soil properties presents challenges to geotechnical engineers. Boundary conditionsare often well defined in other branches of civil engineering, but unlike steel or concrete, the material properties and behavior of soil are difficult to predict due to its variability and limitation oninvestigation. Furthermore, soil exhibits nonlinear (stress-dependent) strength, stiffness, and dilatancy (volume change associated with application of shear stress), making studying soil mechanics all the more difficult.

Water resources engineering

Water resources engineering is concerned with the collection and management of water (as a natural resource). As a discipline it therefore combines hydrology, environmental science, meteorology, geology, conservation, and resource management. This area of civil engineering relates to the prediction and management of both the quality and the quantity of water in both underground (aquifers) and above ground (lakes, rivers, and streams) resources. Water resource engineers analyze and model very small to very large areas of the earth to predict the amount and content of water as it flows into, through, or out of a facility. Although the actual design of the facility may be left to other engineers.

Hydraulic engineering is concerned with the flow and conveyance of fluids, principally water. This area of civil engineering is intimately related to the design of pipelines, water supply network, drainage facilities (including bridges, dams, channels, culverts, levees, storm sewers), and canals. Hydraulic engineers design these facilities using the concepts of fluid pressure, fluid statics, fluid dynamics, and hydraulics, among others.

Structural engineering

Structural engineering is concerned with the structural designand structural analysis of buildings, bridges, towers, flyovers(overpasses), tunnels, off shore structures like oil and gas fields in the sea, aerostructure and other structures. This involves identifying the loads which act upon a structure and the forces and stresses which arise within that structure due to those loads, and then designing the structure to successfully support and resist those loads. The loads can be self weight of the structures, other dead load, live loads, moving (wheel) load, wind load, earthquake load, load from temperature change etc. The structural engineer must design structures to be safe for their users and to successfully fulfill the function they are designed for (to be serviceable). Due to the nature of some loading conditions, sub-disciplines within structural engineering have emerged, including wind engineering and earthquake engineering. Design considerations will include strength, stiffness, and stability of the structure when subjected to loads which may be static, such as furniture or self-weight, or dynamic, such as wind, seismic, crowd or vehicle loads, or transitory, such as temporary construction loads or impact. Other considerations include cost, constructability, safety, aesthetics and sustainability.

Surveying

Surveying is the process by which a surveyor measures certain dimensions that occur on or near the surface of the Earth. Surveying equipment, such as levels and theodolites, are used for accurate measurement of angular deviation, horizontal, vertical and slope distances. With computerisation, electronic distance measurement (EDM), total stations, GPS surveying and laser scanning have to a large extent supplanted traditional instruments. Data collected by survey measurement is converted into a graphical representation of the Earth's surface in the form of a map. This information is then used by civil engineers, contractors and realtors to design from, build on, and trade, respectively. Elements of a structure must be sized and positioned in relation to each other and to site boundaries and adjacent structures. Although surveying is a distinct profession with separate qualifications and licensing arrangements, civil engineers are trained in the basics of surveying and mapping, as well as geographic information systems. Surveyors also lay out the routes of railways, tramway tracks, highways, roads, pipelines and streets as well as position other infrastructure, such as harbors, before construction.

LAND SURVEYING

In the United States, Canada, the United Kingdom and most Commonwealth countries land surveying is considered to be a distinct profession. Land surveyors are not considered to be engineers, and have their own professional associations and licencing requirements. The services of a licenced land surveyor are generally required for boundary surveys (to establish the boundaries of a parcel using its legal description) and subdivision plans (a plot or map based on a survey of a parcel of land, with boundary lines drawn inside the larger parcel to indicate the creation of new boundary lines and roads), both of which are generally referred to as Cadastral surveying.

Construction surveying

Construction surveying is generally performed by specialised technicians. Unlike land surveyors, the resulting plan does not have legal status. Construction surveyors perform the following tasks:

- Surveying existing conditions of the future work site, including topography, existing buildings and infrastructure, and underground infrastructure when possible;

- "lay-out" or "setting-out": placing reference points and markers that will guide the construction of new structures such as roads or buildings;

- Verifying the location of structures during construction;

- As-Built surveying: a survey conducted at the end of the construction project to verify that the work authorized was completed to the specifications set on plans.

Transportation engineering

Transportation engineering is concerned with moving people and goods efficiently, safely, and in a manner conducive to a vibrant community. This involves specifying, designing, constructing, and maintaining transportation infrastructure which includes streets, canals, highways, rail systems, airports, ports, and mass transit. It includes areas such as transportation design, transportation planning, traffic engineering, some aspects of urban engineering, queueing theory, pavement engineering, Intelligent Transportation System (ITS), and infrastructure management.

Forensic engineering

Forensic engineering is the investigation of materials, products, structures or components that fail or do not operate or function as intended, causing personal injury or damage to property. The consequences of failure are dealt with by the law of product liability. The field also deals with retracing processes and procedures leading to accidents in operation of vehicles or machinery. The subject is applied most commonly in civil law cases, although it may be of use in criminal law cases. Generally the purpose of a Forensic engineering investigation is to locate cause or causes of failure with a view to improve performance or life of a component, or to assist a court in determining the facts of an accident. It can also involve investigation of intellectual property claims, especially patents.

Municipal or urban engineering

Municipal engineering is concerned with municipal infrastructure. This involves specifying, designing, constructing, and maintaining streets, sidewalks, water supply networks, sewers, street lighting, municipal solid waste management and disposal, storage depots for various bulk materials used for maintenance and public works (salt, sand, etc.), public parks and bicycle paths. In the case of underground utility networks, it may also include the civil portion (conduits and access chambers) of the local distribution networks of electrical and telecommunications services. It can also include the optimizing of waste collection and bus service networks. Some of these disciplines overlap with other civil engineering specialties, however municipal engineering focuses on the coordination of these infrastructure networks and services, as they are often built simultaneously, and managed by the same municipal authority. Municipal engineers may also design the site civil works for large buildings, industrial plants or campuses (i. e. access roads, parking lots, potable water supply, treatment or pretreatment of waste water, site drainage, etc.)

Control engineering

Control engineering (or control systems engineering) is the branch of civil engineering discipline that applies control theory to design systems with desired behaviors. The practice uses sensors to measure the output performance of the device being controlled (often a vehicle) and those measurements can be used to give feedback to the input actuators that can make corrections toward desired performance. When a device is designed to perform without the need of human inputs for correction it is called automatic control (such as cruise control for regulating a car's speed). Multidisciplinary in nature, control systems engineering activities focus on implementation of control systems mainly derived by mathematical modeling of systems of a diverse range.

FLOODPLAIN AND GROUNDWATER

A floodplain or flood plain is an area of land adjacent to a stream or river that stretches from the banks of its channel to the base of the enclosing valley walls and experiences flooding during periods of high discharge. It includes the floodway, which consists of the stream channel and adjacent areas that actively carry flood flows downstream, and the flood fringe, which are areas inundated by the flood, but which do not experience a strong current. In other words, a floodplain is an area near a river or a stream which floods when the water level reaches flood stage.

FORMATION

Flood plains are made by a meander eroding sideways as they travel downstream. When a river breaks its banks and floods, it leaves behind layers of alluvium (silt). These gradually build up to create the floor of the flood plain. Floodplains generally contain unconsolidated sediments, often extending below the bed of the stream. These are accumulations of sand, gravel, loam, silt, and/or clay, and are often important aquifers, the water drawn from them being pre-filtered compared to the water in the river.

Geologically ancient floodplains are often represented in the landscape by fluvial terraces. These are old floodplains that remain relatively high above the present floodplain and indicate former courses of a stream.

Sections of the Missouri River floodplain taken by the United States Geological Survey show a great variety of material of varying coarseness, the stream bed having been scoured at one place and filled at another by currents and floods of varying swiftness, so that sometimes the deposits are of coarse gravel, sometimes of fine sand or of fine silt. It is probable that any section of such an alluvial plainwould show deposits of a similar character.

The floodplain during its formation is marked by meandering or anastomotic streams, oxbow lakes andbayous, marshes or stagnant pools, and is occasionally completely covered with water. When the drainage system has ceased to act or is entirely diverted for any reason, the floodplain may become a level area of great fertility, similar in appearance to the floor of an old lake. The floodplain differs, however, because it is not altogether flat. It has a gentle slope downstream, and often, for a distance, from the side towards the center.

The floodplain is the natural place for a river to dissipate its energy. Meanders form over the floodplain to slow down the flow of water and when the channel is at capacity the water spills over the floodplain where it is temporarily stored. In terms of flood management the upper part of the floodplain (piedmont zone) is crucial as this is where the flood water control starts. Artificial canalisation of the river here will have a major impact on wider flooding. This is the basis of sustainable flood management.

Ecology

Floodplains can support particularly rich ecosystems, both in quantity and diversity. Tugay forests form an ecosystem associated with floodplains, especially in Central Asia. They are a category of riparianzones or systems. A floodplain can contain 100 or even 1, 000 times as many species as a river. Wetting of the floodplain soil releases an immediate surge of nutrients: those left over from the last flood, and those that result from the rapid decomposition of organic matter that has accumulated since then. Microscopic organisms thrive and larger species enter a rapid breeding cycle. Opportunistic feeders (particularly birds) move in to take advantage. The production of nutrients peaks and falls away quickly; however the surge of new growth endures for some time. This makes floodplains particularly valuable for agriculture.

Interaction with society

Historically, many towns have been built on floodplains, where they are highly susceptible to flooding, for a number of reasons:

- Access to fresh water;
- The fertility of floodplain land for farming;
- Cheap transportation, via rivers and railroads, which often followed rivers;
- Ease of development of flat land

Excluding famines and epidemics, some of the worst natural disasters in history (measured by fatalities) have been river floods, particularly in the Yellow River in China – see list of deadliest floods. The worst of these, and the worst natural disaster (excluding famine and epidemics) were the 1931 China floods, estimated to have killed millions. This had been preceded by the 1887 Yellow River flood, which killed around one million people, and is the second-worst natural disaster in history.

The extent of floodplain inundation depends in part on the flood magnitude, defined by the return period.

In the United States the Federal Emergency Management Agency (FEMA) manages the National Flood Insurance Program (NFIP). The NFIP offers insurance to properties located within a flood prone area, as defined by the Flood Insurance Rate Map (FIRM), which depicts various flood risks for a community. The FIRM typically focuses on delineation of the 100-year flood inundation area, also known within the NFIP as the Special Flood Hazard Area.

Where a detailed study of a waterway has been done, the 100-year floodplain will also include the floodway, the critical portion of the floodplain which includes the stream channel and any adjacent areas that must be kept free of encroachments that might block flood flows or restrict storage of flood waters. Another commonly encountered term is the Special Flood Hazard Area, which is any area subject to inundation by the 100-year flood. A problem is that any alteration of the watershed upstream of the point in question can potentially affect the ability of the watershed to handle water, and thus potentially affects the levels of the periodic floods. A large shopping center and parking lot, for example, may raise the levels of the 5-year, 100-year, and other floods, but the maps are rarely adjusted, and are frequently rendered obsolete by subsequent development.

In order for flood-prone property to qualify for government-subsidized insurance, a local community must adopt an ordinance that protects the floodway and requires that new residential structures built in Special Flood Hazard Areas be elevated to at least the level of the 100-year flood. Commercial structures can be elevated or flood proofed to or above this level. In some areas without detailed study information, structures may be required to be elevated to at least two feet above the surrounding grade. Many State and local governments have, in addition, adopted floodplain construction regulations which are more restrictive than those mandated by the NFIP. The U. S. government also sponsors flood hazard mitigation efforts to reduce flood impacts. The Hazard Mitigation Program is one funding source for mitigation projects. A number of whole towns such as English, Indiana, have been completely relocated to remove them from the floodplain. Other smaller-scale mitigation efforts include acquiring and demolishing flood-prone buildings or flood-proofing them.

In some tropical floodplain areas such as the Inner Niger Delta of Mali, annual flooding events are a natural part of the local ecology and rural economy, allowing for the raising of crops through recessional agriculture. But in Bangladesh, which occupies the Ganges Delta, the advantages provided by the richness of the alluvial soil of floodplains are severely offset by frequent floods brought on by cyclones and annualmonsoon rains, which cause severe economic disruption and loss of human life in this densely populated region.

Groundwater

Groundwater is the water located beneath Earth's surface in soil pore spaces and in the fractures ofrock formations. A unit of rock or an unconsolidated deposit is called an aquifer when it can yield a usable quantity of water. The depth at which soil pore spaces or fractures and voids in rock become completely saturated with water is called the water table. Groundwater is recharged from, and eventually flows to, the surface naturally; natural discharge often occurs at springs and seeps, and can form oases or wetlands. Groundwater is also often withdrawn for agricultural, municipal, andindustrial use by constructing and operating extraction wells. The study of the distribution and movement of groundwater is hydrogeology, also called groundwater hydrology.

Typically, groundwater is thought of as liquid water flowing through shallow aquifers, but, in the technical sense, it can also include soil moisture, permafrost (frozen soil), immobile water in very low permeability bedrock, and deep geothermal or oil formation water. Groundwater is hypothesized to provide lubrication that can possibly influence the movement of faults. It is likely that much of Earth's subsurface contains some water, which may be mixed with other fluids in some instances. Groundwater may not be confined only to Earth. The formation of some of the landforms observed on Mars may have been influenced by groundwater. There is also evidence that liquid water may also exist in the subsurface of Jupiter's moon Europa.

Aquifers

An aquifer is a layer of porous substrate that contains and transmits groundwater. When water can flow directly between the surface and the saturated zone of an aquifer, the aquifer is unconfined. The deeper parts of unconfined aquifers are usually more saturated since gravity causes water to flow downward.

The upper level of this saturated layer of an unconfined aquifer is called the water table or phreatic surface. Below the water table, where in general all pore spaces are saturated with water, is the phreatic zone.

Substrate with low porosity that permits limited transmission of groundwater is known as an aquitard. Anaquiclude is a substrate with porosity that is so low it is virtually impermeable to groundwater.

A confined aquifer is an aquifer that is overlain by a relatively impermeable layer of rock or substrate such as an aquiclude or aquitard. If a confined aquifer follows a downward grade from its recharge zone, groundwater can become pressurized as it flows. This can create artesian wells that flow freely without the need of a pump and rise to a higher elevation than the static water table at the above, unconfined, aquifer.

The characteristics of aquifers vary with the geology and structure of the substrate and topography in which they occur. In general, the more productive aquifers occur in sedimentary geologic formations. By comparison, weathered and fractured crystalline rocks yield smaller quantities of groundwater in many environments. Unconsolidated to poorly cemented alluvial materials that have accumulated as valley-filling sediments in major river valleys and geologically subsiding structural basins are included among the most productive sources of groundwater.

The high specific heat capacity of water and the insulating effect of soil and rock can mitigate the effects of climate and maintain groundwater at a relatively steady temperature. In some places where groundwater temperatures are maintained by this effect at about 10 °C (50 °F), groundwater can be used for controlling the temperature inside structures at the surface. For example, during hot weather relatively cool groundwater can be pumped through radiators in a home and then returned to the ground in another well. During cold seasons, because it is relatively warm, the water can be used in the same way as a source of heat for heat pumps that is much more efficient than using air.

Groundwater is often cheaper, more convenient and less vulnerable to pollution than surface water. Therefore, it is commonly used for public water supplies. Groundwater provides the largest source of usable water storage in the United States. Underground reservoirs contain far more water than the capacity of all surface reservoirs and lakes, including the Great Lakes. Many municipal water supplies are derived solely from groundwater.

The volume of groundwater in an aquifer can be estimated by measuring water levels in local wells and by examining geologic records from well-drilling to determine the extent, depth and thickness of water-bearing sediments and rocks. Before an investment is made in production wells, test wells may be drilled to measure the depths at which water is encountered and collect samples of soils, rock and water for laboratory analyses. Pumping tests can be performed in test wells to determine flow characteristics of the aquifer.

Polluted ground water is less visible, but more difficult to clean up, than pollution in rivers and lakes. Ground water pollution most often results from improper disposal of wastes on land. Major sources include industrial and household chemicals and garbage landfills, industrial waste lagoons, tailings and process wastewater from mines, oil field brine pits, leaking underground oil storage tanks and pipelines, sewage sludge and septic systems. Polluted groundwater is mapped by sampling soils and groundwater near suspected or known sources of pollution, to determine the extent of the pollution, and to aid in the design of groundwater remediation systems. Preventing groundwater pollution near potential sources such as landfills requires lining the bottom of a landfill with watertight materials, collecting any leachate with drains, and keeping rainwater off any potential contaminants, along with regular monitoring of nearby groundwater to verify that contaminants have not leaked into the groundwater.

The danger of pollution of municipal supplies is minimized by locating wells in areas of deep groundwater and impermeable soils, and careful testing and monitoring of the aquifer and nearby potential pollution sources.

Water cycle

Groundwater makes up about twenty percent of the world's fresh water supply, which is about 0. 61% of the entire world's water, including oceans and permanent ice. Global groundwater storage is roughly equal to the total amount of freshwater stored in the snow and ice pack, including the north and south poles. This makes it an important resource that can act as a natural storage that can buffer against shortages of surface water, as in during times of drought.

Groundwater is naturally replenished by surface water from precipitation, streams, and rivers when this recharge reaches the water table.

Groundwater can be a long-term 'reservoir' of the natural water cycle (with residence times from days to millennia), as opposed to short-term water reservoirs like the atmosphere and fresh surface water (which have residence times from minutes to years). The figure shows how deep groundwater (which is quite distant from the surface recharge) can take a very long time to complete its natural cycle.

The Great Artesian Basin in central and eastern Australia is one of the largest confined aquifer systems in the world, extending for almost 2 million km2. By analysing the trace elements in water sourced from deep underground, hydrogeologists have been able to determine that water extracted from these aquifers can be more than 1 million years old.

By comparing the age of groundwater obtained from different parts of the Great Artesian Basin, hydrogeologists have found it increases in age across the basin. Where water recharges the aquifers along the Eastern Divide, ages are young. As groundwater flows westward across the continent, it increases in age, with the oldest groundwater occurring in the western parts. This means that in order to have travelled almost 1000 km from the source of recharge in 1 million years, the groundwater flowing through the Great Artesian Basin travels at an average rate of about 1 metre per year.

Recent research has demonstrated that evaporation of groundwater can play a significant role in the local water cycle, especially in arid regions. Scientists in Saudi Arabia have proposed plans to recapture and recycle this evaporative moisture for crop irrigation. In the opposite photo, a 50-centimeter-square reflective carpet, made of small adjacent plastic cones, was placed in a plant-free dry desert area for five months, without rain or irrigation. It managed to capture and condense enough ground vapor to bring to life naturally buried seeds underneath it, with a green area of about 10% of the carpet area. It is expected that, if seeds were put down before placing this carpet, a much wider area would become green.

ISSUES

Subsidence

Subsidence occurs when too much water is pumped out from underground, deflating the space below the above-surface, and thus causing the ground to collapse. The result can look like craters on plots of land. This occurs because, in its natural equilibrium state, the hydraulic pressure of groundwater in the pore spaces of the aquifer and the aquitard supports some of the weight of the overlying sediments. When groundwater is removed from aquifers by excessive pumping, pore pressures in the aquifer drop and compression of the aquifer may occur. This compression may be partially recoverable if pressures rebound, but much of it is not. When the aquifer gets compressed, it may cause land subsidence, a drop in the ground surface. The city of New Orleans, Louisiana is actually below sea level today, and its subsidence is partly caused by removal of groundwater from the various aquifer/aquitard systems beneath it. In the first half of the 20th century, the city of San Jose, California dropped 13 feet from land subsidence caused by overpumping; this subsidence has been halted with improved groundwater management.

Seawater intrusion

In general, in very humid or undeveloped regions, the shape of the water table mimics the slope of the surface. The recharge zone of an aquifer near the seacoast is likely to be inland, often at considerable distance. In these coastal areas, a lowered water table may induce sea water to reverse the flow toward the land. Sea water moving inland is called a saltwater intrusion. In alternative fashion, salt from mineralbeds may leach into the groundwater of its own accord.

IGNEOUS ROCK 21

Igneous rock (derived from the Latin word ignis meaning fire) is one of the three main rock types, the others being sedimentary and metamorphic. Igneous rock is formed through the cooling and solidification of magma or lava. Igneous rock may form with or without crystallization, either below the surface as intrusive (plutonic) rocks or on the surface as extrusive (volcanic) rocks. This magma can be derived from partial melts of pre-existing rocks in either a planet's mantle or crust. Typically, the melting is caused by one or more of three processes: an increase in temperature, a decrease in pressure, or a change in composition. Over 700 types of igneous rocks have been described, most of them having formed beneath the surface of Earth's crust.

GEOLOGICAL SIGNIFICANCE

Igneous and metamorphic rocks make up 90–95% of the top 16 km of the Earth's crust by volume.

Igneous rocks are geologically important because:

⊙ Their minerals and global chemistry give information about the composition of the mantle, from which some igneous rocks are extracted, and the temperature and pressure conditions that allowed this extraction, and/or of other pre-existing rock that melted;

⊙ Their absolute ages can be obtained from various forms of radiometric dating and thus can be compared to adjacent geological strata, allowing a time sequence of events;

⊙ Their features are usually characteristic of a specific tectonic environment, allowing tectonic reconstitutions;

⊙ in some special circumstances they host important mineral deposits (ores): for example, tungsten, tin, and uranium are commonly associated with granites and diorites, whereas ores of chromium and platinum are commonly associated with gabbros.

Morphology and setting

In terms of modes of occurrence, igneous rocks can be either intrusive (plutonic), extrusive (volcanic) or hypabyssal.

Intrusive

Intrusive igneous rocks are formed from magma that cools and solidifies within the crust of a planet, surrounded by pre-existing rock (called country rock); the magma cools slowly and, as a result, these rocks are coarse grained. The mineral grains in such rocks can generally be identified with the naked eye. Intrusive rocks can also be classified according to the shape and size of the intrusive body and its relation to the other formations into which it intrudes. Typical intrusive formations are batholiths, stocks, laccoliths, sills and dikes. When the magma solidifies within the earth's crust, it cools slowly forming coarse textured rocks, such as granite, gabbro, or diorite.

The central cores of major mountain ranges consist of intrusive igneous rocks, usually granite. When exposed by erosion, these cores (called batholiths) may occupy huge areas of the Earth's surface.

Coarse grained intrusive igneous rocks that form at depth within the crust are termed as abyssal; intrusive igneous rocks that form near the surface are termed hypabyssal.

Extrusive

Extrusive igneous rocks, also known as volcanic rocks, are formed at the crust's surface as a result of the partial melting of rocks within the mantle and crust. Extrusive igneous rocks cool and solidify quicker than intrusive igneous rocks. They are formed by the cooling of molten magma on the earth's surface. The magma, which is brought to the surface through fissures or volcanic eruptions, solidifies at a faster rate. Hence such rocks are smooth, crystalline and fine grained. Basalt is a common extrusive igneous rock and forms lava flows, lava sheets and lava plateaus. Some kinds of basalt solidify to form long polygonal columns. The Giant's Causeway found in Antrim, Northern Ireland is an example.

The melted rock, with or without suspended crystals and gas bubbles, is called magma. It rises because it is less dense than the rock from which it was created. When magma reaches the surface from beneath water or air, it is called lava. Eruptions of volcanoes into air are termed subaerial, whereas those occurring underneath the ocean are termed submarine. Black smokers and mid-ocean ridge basalt are examples of submarine volcanic activity.

The volume of extrusive rock erupted annually by volcanoes varies with plate tectonic setting. Extrusive rock is produced in the following proportions:

- ◉ Divergent boundary: 73%

- ◉ Convergent boundary (subduction zone): 15%

- ◉ Hotspot: 12%.

Magma that erupts from a volcano behaves according to its viscosity, determined by temperature, composition, and crystal content. High-temperature magma, most of which is basaltic in composition, behaves in a manner similar to thick oil and, as it cools, treacle. Long, thin basalt flows with pahoehoe surfaces are common. Intermediate composition magma, such as andesite, tends to form cinder cones of intermingled ash, tuff and lava, and may have a viscosity similar to thick, cold molasses or even rubber when erupted. Felsic magma, such as rhyolite, is usually erupted at low temperature and is up to 10,000 times as viscous as basalt. Volcanoes with rhyolitic magma commonly erupt explosively, and rhyolitic lava flows are typically of limited extent and have steep margins, because the magma is so viscous.

Felsic and intermediate magmas that erupt often do so violently, with explosions driven by the release of dissolved gases—typically water vapour, but also carbon dioxide. Explosively erupted pyroclastic material is called tephra and includes tuff, agglomerate and ignimbrite. Fine volcanic ash is also erupted and forms ash tuff deposits, which can often cover vast areas.

Because lava cools and crystallizes rapidly, it is fine grained. If the cooling has been so rapid as to prevent the formation of even small crystals after extrusion, the resulting rock may be mostly glass (such as the rock obsidian). If the cooling of the lava happened more slowly, the rocks would be coarse-grained.

Because the minerals are mostly fine-grained, it is much more difficult to distinguish between the different types of extrusive igneous rocks than between different types of intrusive igneous rocks. Generally, the mineral constituents of fine-grained extrusive igneous rocks can only be determined by examination of thin sections of the rock under a microscope, so only an approximate classification can usually be made in the field.

Hypabyssal

Hypabyssal igneous rocks are formed at a depth in between the plutonic and volcanic rocks. These are formed due to cooling and resultant solidification of rising magma just beneath the earth surface. Hypabyssal rocks are less common than plutonic or volcanic rocks and often form dikes, sills, laccoliths, lopoliths, or phacoliths.

Classification

Igneous rocks are classified according to mode of occurrence, texture, mineralogy, chemical composition, and the geometry of the igneous body.

The classification of the many types of different igneous rocks can provide us with important information about the conditions under which they formed. Two important variables used for the classification of igneous rocks are particle size, which largely depends on the cooling history, and the mineral composition of the rock. Feldspars, quartz or feldspathoids, olivines, pyroxenes, amphiboles, and micas are all important minerals in the formation of almost all igneous rocks, and they are basic to the classification of these rocks. All other minerals present are regarded as nonessential in almost all igneous rocks and are called accessory minerals. Types of igneous rocks with other essential minerals are very rare, and these rare rocks include those with essential carbonates.

In a simplified classification, igneous rock types are separated on the basis of the type of feldspar present, the presence or absence of quartz, and in rocks with no feldspar or quartz, the type of iron or magnesium minerals present. Rocks containing quartz (silica in composition) are silica-oversaturated. Rocks with feldspathoids are silica-undersaturated, because feldspathoids cannot coexist in a stable association with quartz.

Igneous rocks that have crystals large enough to be seen by the naked eye are called phaneritic; those with crystals too small to be seen are called aphanitic. Generally speaking, phaneritic implies an intrusive origin; aphanitic an extrusive one.

An igneous rock with larger, clearly discernible crystals embedded in a finer-grained matrix is termed porphyry. Porphyritic texture develops when some of the crystals grow to considerable size before the main mass of the magma crystallizes as finer-grained, uniform material.

Igneous rocks are classified on the basis of texture and composition. Texture refers to the size, shape, and arrangement of the mineral grains or crystals of which the rock is composed.

Texture

Texture is an important criterion for the naming of volcanic rocks. The texture of volcanic rocks, including the size, shape, orientation, and distribution of mineral grains and the intergrain relationships, will determine whether the rock is termed a tuff, a pyroclastic lava or a simple lava.

However, the texture is only a subordinate part of classifying volcanic rocks, as most often there needs to be chemical information gleaned from rocks with extremely fine-grained groundmass or from airfall tuffs, which may be formed from volcanic ash.

Textural criteria are less critical in classifying intrusive rocks where the majority of minerals will be visible to the naked eye or at least using a hand lens, magnifying glass or microscope. Plutonic rocks also tend to be less texturally varied and less prone to gaining structural fabrics. Textural terms can be used to differentiate different intrusive phases of large plutons, for instance porphyritic margins to large intrusive bodies, porphyry stocks and subvolcanic dikes (apophyses). Mineralogical classification is most often used to classify plutonic rocks. Chemical classifications are preferred to classify volcanic rocks, with phenocryst species used as a prefix, e.g. "olivine-bearing picrite" or "orthoclase-phyric rhyolite".

Chemical classification and Petrology

Igneous rocks can be classified according to chemical or mineralogical parameters.

Chemical: total alkali-silica content (TAS diagram) for volcanic rock classification used when modal or mineralogic data is unavailable:

◉ Felsic igneous rocks containing a high silica content, greater than 63% SiO_2 (examples granite and rhyolite)

◉ Intermediate igneous rocks containing between 52 – 63% SiO_2 (example andesite and dacite)

◉ Mafic igneous rocks have low silica 45 – 52% and typically high iron – magnesium content (example gabbro and basalt)

◉ Ultramafic rock igneous rocks with less than 45% silica. (examples picrite, komatiite and peridotite)

◉ Alkalic igneous rocks with 5 – 15% alkali ($K_2O + Na_2O$) content or with a molar ratio of alkali to silica greater than 1:6. (examples phonolite and trachyte)

Chemical classification also extends to differentiating rocks that are chemically similar according to the TAS diagram, for instance;

◉ Ultrapotassic; rocks containing molar K_2O/Na_2O >3

◉ Peralkaline; rocks containing molar ($K_2O + Na_2O$)/ Al_2O_3 >1

◉ Peraluminous; rocks containing molar ($K_2O + Na_2O$)/ Al_2O_3 <1

An idealized mineralogy (the normative mineralogy) can be calculated from the chemical composition, and the calculation is useful for rocks too fine-grained or too altered for identification of minerals that crystallized from the melt. For instance, normative quartz classifies a rock as silica-oversaturated; an example is rhyolite. In an older terminology, silica oversaturated rocks were called silicic or acidic where the SiO_2 was greater than 66% and the family term quartzolite was applied to the most silicic. A normative feldspathoid classifies a rock as silica-undersaturated; an example is nephelinite.

History of classification

In 1902, a group of American petrographers proposed that all existing classifications of igneous rocks should be discarded and replaced by a "quantitative" classification based on chemical analysis. They showed how vague, and often unscientific, much of the existing terminology was and argued that as the chemical composition of an igneous rock was its most fundamental characteristic, it should be elevated to prime position. Geological occurrence, structure, mineralogical constitution—the hitherto accepted criteria for the discrimination of rock species—were relegated to the background. The completed rock analysis is first to be interpreted in terms of the rock-forming minerals which might be expected to be formed when the magma crystallizes, e.g., quartz feldspars, olivine, akermannite, Feldspathoids, magnetite, corundum, and so on, and the rocks are divided into groups strictly according to the relative proportion of these minerals to one another.

Mineralogical classification

For volcanic rocks, mineralogy is important in classifying and naming lavas. The most important criterion is the phenocryst species, followed by the groundmass mineralogy. Often, where the groundmass is aphanitic, chemical classification must be used to properly identify a volcanic rock.

Mineralogic contents – felsic versus mafic

- ⊙ Felsic rock, highest content of silicon, with predominance of quartz, alkali feldspar and/or feldspathoids: the felsic minerals; these rocks (e.g., granite, rhyolite) are usually light coloured, and have low density.

- ⊙ Mafic rock, lesser content of silicon relative to felsic rocks, with predominance of mafic minerals pyroxenes, olivines and calcic plagioclase; these rocks (example, basalt, gabbro) are usually dark coloured, and have a higher density than felsic rocks.

- ⊙ Ultramafic rock, lowest content of silicon, with more than 90% of mafic minerals (e.g., dunite).

For intrusive, plutonic and usually phaneritic igneous rocks (where all minerals are visible at least via microscope), the mineralogy is used to classify the rock. This usually occurs on ternary diagrams, where the relative proportions of three minerals are used to classify the rock.

Example of classification

Granite is an igneous intrusive rock (crystallized at depth), with felsic composition (rich in silica and predominately quartz plus potassium-rich feldspar plus sodium-rich plagioclase) and phaneritic, subeuhedral texture (minerals are visible to the unaided eye and commonly some of them retain original crystallographic shapes).

Magma origination

The Earth's crust averages about 35 kilometers thick under the continents, but averages only some 7–10 kilometers beneath the oceans. The continental crust is composed primarily of sedimentary rocks resting on a crystalline basement formed of a great variety of metamorphic and igneous rocks, including granulite and granite. Oceanic crust is composed primarily of basalt and gabbro. Both continental and oceanic crust rest on peridotite of the mantle.

Rocks may melt in response to a decrease in pressure, to a change in composition (such as an addition of water), to an increase in temperature, or to a combination of these processes.

Other mechanisms, such as melting from a meteorite impact, are less important today, but impacts during the accretion of the Earth led to extensive melting, and the outer several hundred kilometers of our early Earth was probably an ocean of magma. Impacts of large meteorites in the last few hundred million years have been proposed as one mechanism responsible for the extensive basalt magmatism of several large igneous provinces.

Decompression

Decompression melting occurs because of a decrease in pressure.

The solidus temperatures of most rocks (the temperatures below which they are completely solid) increase with increasing pressure in the absence of water. Peridotite at depth in the Earth's mantle may be hotter than its solidus temperature at some shallower level. If such rock rises during the convection of solid mantle, it will cool slightly as it expands in an adiabatic process, but the cooling is only about 0.3 °C per kilometer. Experimental studies of appropriate peridotite samples document that the solidus temperatures increase by 3 °C to 4 °C per kilometer. If the rock rises far enough, it will begin to melt. Melt droplets can coalesce into larger volumes and be intruded upwards. This process of melting from the upward movement of solid mantle is critical in the evolution of the Earth.

Decompression melting creates the ocean crust at mid-ocean ridges. It also causes volcanism in intraplate regions, such as Europe, Africa and the Pacific sea floor. There, it is variously attributed either to the rise of mantle plumes (the "Plume hypothesis") or to intraplate extension (the "Plate hypothesis").

Effects of water and carbon dioxide

The change of rock composition most responsible for the creation of magma is the addition of water. Water lowers the solidus temperature of rocks at a given pressure. For example, at a depth of about 100 kilometers, peridotite begins to melt near 800 °C in the presence of excess water, but near or above about 1,500 °C in the absence of water. Water is driven out of the oceanic lithosphere in subduction zones, and it causes melting in the overlying mantle. Hydrous magmas composed of basalt and andesite are produced directly and indirectly as results of dehydration during the subduction process. Such magmas, and those derived from them, build up island arcs such as those in the Pacific Ring of Fire. These magmas form rocks of the calc-alkaline series, an important part of the continental crust.

The addition of carbon dioxide is relatively a much less important cause of magma formation than the addition of water, but genesis of some silica-undersaturated magmas has been attributed to the dominance of carbon dioxide over water in their mantle source regions. In the presence of carbon dioxide, experiments document that the peridotite solidus temperature decreases by about 200 °C in a narrow pressure interval at pressures corresponding to a depth of about 70 km. At greater depths, carbon dioxide can have more effect: at depths to about 200 km, the temperatures of initial melting of a carbonated peridotite composition were determined to be 450 °C to 600 °C lower than for the same composition with no carbon dioxide. Magmas of rock types such as nephelinite, carbonatite, and kimberlite are among those that may be generated following an influx of carbon dioxide into mantle at depths greater than about 70 km.

Temperature increase

Increase in temperature is the most typical mechanism for formation of magma within continental crust. Such temperature increases can occur because of the upward intrusion of magma from the mantle. Temperatures can also exceed the solidus of a crustal rock in continental crust thickened by compression at a plate boundary. The plate boundary between the Indian and Asian continental masses provides a well-studied example, as the Tibetan Plateau just north of the boundary has crust about 80 kilometers thick, roughly twice the thickness of normal continental crust. Studies of electrical resistivity deduced from magnetotelluric data have detected a layer that appears to contain silicate melt and that stretches for at least 1,000 kilometers within the middle crust along the southern margin of the Tibetan Plateau. Granite and rhyolite are types of igneous rock commonly interpreted as products of the melting of continental crust because of increases in temperature. Temperature increases also may contribute to the melting of lithosphere dragged down in a subduction zone.

Magma evolution

Most magmas only entirely melt for small parts of their histories. More typically, they are mixes of melt and crystals, and sometimes also of gas bubbles. Melt, crystals, and bubbles usually have different densities, and so they can separate as magmas evolve.

As magma cools, minerals typically crystallize from the melt at different temperatures (fractional crystallization). As minerals crystallize, the composition of the residual melt typically changes. If crystals separate from the melt, then the residual melt will differ in composition from the parent magma. For instance, a magma of gabbroic composition can produce a residual melt of granitic composition if early formed crystals are separated from the magma. Gabbro may have a liquidus temperature near 1,200 °C, and the derivative granite-composition melt may have a liquidus temperature as low as about 700 °C. Incompatible elements are concentrated in the last residues of magma during fractional crystallization and in the first melts produced during partial melting: either process can form the magma that crystallizes to pegmatite, a rock type commonly enriched in incompatible elements. Bowen's reaction series is important for understanding the idealised sequence of fractional crystallisation of a magma.

Magma composition can be determined by processes other than partial melting and fractional crystallization. For instance, magmas commonly interact with rocks they intrude, both by melting those rocks and by reacting with them. Magmas of different compositions can mix with one another. In rare cases, melts can separate into two immiscible melts of contrasting compositions.

There are relatively few minerals that are important in the formation of common igneous rocks, because the magma from which the minerals crystallize is rich in only certain elements: silicon, oxygen, aluminium, sodium, potassium, calcium, iron, and magnesium. These are the elements that combine to form the silicate minerals, which account for over ninety percent of all igneous rocks. The chemistry of igneous rocks is expressed differently for major and minor elements and for trace elements. Contents of major and minor elements are conventionally expressed as weight percent oxides (e.g., 51% SiO_2, and 1.50% TiO_2). Abundances of trace elements are conventionally expressed as parts per million by weight (e.g., 420 ppm Ni, and 5.1 ppm Sm). The term "trace element" is typically used for elements present in most rocks at abundances less than 100 ppm or so, but some trace elements may be present in some rocks at abundances exceeding 1,000 ppm. The diversity of rock compositions has been defined by a huge mass of analytical data—over 230,000 rock analyses can be accessed on the web through a site sponsored by the U. S. National Science Foundation.

Etymology

The word "igneous" is derived from the Latin ignis, meaning "of fire". Volcanic rocks are named after Vulcan, the Roman name for the god of fire. Intrusive rocks are also called "plutonic" rocks, named after Pluto, the Roman god of the underworld.

Igneous rocks

The second type of rock we'll look at is igneous rock.

Formation

The inside of the Earth is very hot - hot enough to melt rocks. Molten (liquid) rock forms when rocks melt. The molten rock is called magma. When the magma cools and solidifies, a type of rock called igneous rock forms.

What are they like?

Igneous rocks contain randomly arranged interlocking crystals. The size of the crystals depends on how quickly the molten magma solidified. The more slowly the magma cools, the bigger the crystals.

You may have done an experiment at school with a substance called salol. If molten salol cools slowly, you get big crystals. If it cools quickly, you get small crystals.

CHAPTER

METAMORPHISM AND TAMORPHIC ROCKS

22

Metamorphism is the change of minerals or geologic texture (distinct arrangement of minerals) in pre-existing rocks (protoliths), without the protolith melting into liquid magma (a solid-state change). The change occurs primarily due to heat, pressure, and the introduction of chemically active fluids.

The chemical components and crystal structures of the minerals making up the rock may change even though the rock remains a solid. Changes at or just beneath Earth's surface due to weathering and/or diagenesis are not classified as metamorphism. Metamorphism typically occurs between diagenesis 200°C, and melting 850°C.

Three types of metamorphism exist: contact, dynamic, and regional. Metamorphism produced with increasing pressure and temperature conditions is known as prograde metamorphism. Conversely, decreasing temperatures and pressure characterize retrograde metamorphism.

SOLID STATE RECRYSTALIZATION AND NEOCRYSTALIZATION

Metamorphic rocks can change without melting. Heat causes atomic bonds to break, and the atoms move and form new bonds with other atoms, creating new minerals with different chemical components or crystalline structures (neocrystallization), or enabling recrystallization. When pressure is applied, somewhat flattened grains that orient in the same direction have a more stable configuration.

Limits of metamorphism

The temperature lower limit of metamorphism is considered to be 100 – 200 °C, to exclude diagenetic changes, due to compaction, which result in sedimentary rocks. There is no agreement on a pressure lower limit. Some workers argue that changes in atmospheric pressures are not metamorphic, but some types of metamorphism can occur at extremely low pressures.

The upper boundary of metamorphic conditions is related to the onset of melting processes in the rock. The maximum temperature for metamorphism is typically 700 – 900 °C, depending on the pressure and on the composition of the rock. Migmatites are rocks formed at this upper limit, which contain pods and veins of material that has started to melt but has not fully segregated from the refractory residue. Since the 1980s it has been recognized that, rarely, rocks are dry enough and of a refractory enough composition to record without melting "ultra-high" metamorphic temperatures of 900 – 1100 °C.

Kinds of metamorphism

Regional metamorphism

Regional or Barrovian metamorphism covers large areas of continental crust typically associated with mountain ranges, particularly those associated with convergent tectonic plates or the roots of previously eroded mountains. Conditions producing widespread regionally metamorphosed rocks occur during an orogenic event. The collision of two continental plates or island arcs with continental plates produce the extreme compressional forces required for the metamorphic changes typical of regional metamorphism. These orogenic mountains are later eroded, exposing the intensely deformed rocks typical of their cores. The conditions within the subducting slab as it plunges toward the mantle in a subduction zone also produce regional metamorphic effects, characterized by paired metamorphic belts. The techniques of structural geology are used to unravel the collisional history and determine the forces involved. Regional metamorphism can be described and classified into metamorphic facies or metamorphic zones of temperature/pressure conditions throughout the orogenic terrane.

Contact (thermal) metamorphism

Contact metamorphism occurs typically around intrusive igneous rocks as a result of the temperature increase caused by the intrusion of magma into cooler country rock. The area surrounding the intrusion where the contact metamorphism effects are present is called the metamorphic aureole. Contact metamorphic rocks are usually known as hornfels. Rocks formed by contact metamorphism may not present signs of strong deformation and are often fine-grained.

Contact metamorphism is greater adjacent to the intrusion and dissipates with distance from the contact. The size of the aureole depends on the heat of the intrusion, its size, and the temperature difference with the wall rocks. Dikes generally have small aureoles with minimal metamorphism whereas large ultramafic intrusions can have significantly thick and well-developed contact metamorphism.

The metamorphic grade of an aureole is measured by the peak metamorphic mineral which forms in the aureole. This is usually related to the metamorphic temperatures of pelitic or alumonisilicate rocks and the minerals they form.The metamorphic grades of aureoles are andalusite hornfels, sillimanite hornfels, pyroxene hornfels.

Magmatic fluids coming from the intrusive rock may also take part in the metamorphic reactions. Extensive addition of magmatic fluids can significantly modify the chemistry of the affected rocks. In this case the metamorphism grades into metasomatism. If the intruded rock is rich in carbonate the result is a skarn. Fluorine-rich magmatic waters which leave a cooling granite may often form greisens within and adjacent to the contact of the granite. Metasomatic altered aureoles can localize the deposition of metallic ore minerals and thus are of economic interest.

Hydrothermal metamorphism

Hydrothermal metamorphism is the result of the interaction of a rock with a high-temperature fluid of variable composition. The difference in composition between existing rock and the invading fluid triggers a set of metamorphic and metasomatic reactions. The hydrothermal fluid may be magmatic (originate in an intruding magma), circulating groundwater, or ocean water. Convective circulation of hydrothermal fluids in the ocean floor basalts produces extensive hydrothermal metamorphism adjacent to spreading centers and other submarine volcanic areas. The fluids eventually escape through vents in the ocean floor known as black smokers. The patterns of this hydrothermal alteration is used as a guide in the search for deposits of valuable metal ores.

Shock metamorphism

This kind of metamorphism occurs when either an extraterrestrial object (a meteorite for instance) collides with the Earth's surface or during an extremely violent volcanic eruption. Impact metamorphism is, therefore, characterized by ultrahigh pressure conditions and low temperature. The resulting minerals (such as SiO_2 polymorphs coesite and stishovite) and textures are characteristic of these conditions.

Dynamic metamorphism

Dynamic metamorphism is associated with zones of high to moderate strain such as fault zones. Cataclasis, crushing and grinding of rocks into angular fragments, occurs in dynamic metamorphic zones, giving cataclastic texture.

The textures of dynamic metamorphic zones are dependent on the depth at which they were formed, as the temperature and confining pressure determine the deformation mechanisms which predominate. Within depths less than 5 km, dynamic metamor-phism is not often produced because the confining pressure is too low to produce frictional heat. Instead, a zone of breccia or cataclasite is formed, with the rock milled and broken into random fragments. This generally forms a mélange. At depth, the angular breccias transit into a ductile shear texture and into mylonite zones.

Within the depth range of 5–10 km pseudotachylite is formed, as the confining pressure is enough to prevent brecciation and milling and thus energy is focused into discrete fault planes. Frictional heating in this case may melt the rock to form pseudotachylite glass.

Within the depth range of 10–20 km, deformation is governed by ductile deformation conditions and hence frictional heating is dispersed throughout shear zones, resulting in a weaker thermal imprint and distributed deformation. Here, deformation forms mylonite, with dynamothermal metamorphism observed rarely as the growth of porphyroblasts in mylonite zones.

Overthrusting may juxtapose hot lower crustal rocks against cooler mid and upper crust blocks, resulting in conductive heat transfer and localised contact metamorphism of the cooler blocks adjacent to the hotter blocks, and often retrograde metamorphism in the hotter blocks. The metamorphic assemblages in this case are diagnostic of the depth and temperature and the throw of the fault and can also be dated to give an age of the thrusting.

- Metamorphism: refers to changes in rock texture or mineralogy.

- Metasomatism: means a change in rock composition resulting from diffusion or fluid influx. For example, SiO2-rich fluids can infiltrate ultramafic rocks, which normally have rather little silica, dramatically changing the bulk composition, and stabilizing a new group of high- SiO2 minerals.

- Metamorphic petrology: the study of how almost the entire solid earth works, using chemistry and physics to interpret textures and compositions of minerals; petrologists use theory, experiments, laboratory study, and field study to achieve this goal

- Theory is based on chemistry and physics, particularly thermodynamics, kinetics, and material science

- Experiments use high-P, high-T equipment, such as laser-heated diamond-anvil cells to mimic Earth's interior

- Laboratory investigations include crystal structure and orientation studies by diffraction of x-rays, electrons, and neutrons; texture studies by electron microscopy; and compositional studies using electron microprobe

- Field studies involve mapping, field interpretation of structure and petrology, and collection of samples:

- Metamorphic grade refers qualitatively to the temperature experienced by the rock. The determination of metamorphic grade is made using mineral assemblages, mineral composi-tions, and/or grain sizes

Mineralogical Changes

- Neomineralization/neocrystallization: formation of new minerals (e.g., the appearance of garnet in a rock that lacked garnet).

- Recrystallization: rearrangement of crystal mass or crystal defects (including grain boundaries).

  - If accompanied by deformation, recrystallization can produce a crystallographic preferred orientation, lineation, foliation, and porphyroclasts.

  - If no phases appear or disappear during the recrystalli-zation, only grain growth is occurring; this coarsening is driven by the free energy of grain boundaries, which is the energy resulting from two crystals being in contact along an imperfect boundary.

Metamorphism, mineralogical and structural adjustments of solid rocks to physical and chemical conditions differing from those under which the rocks originally formed. Changes produced by surface conditions such as compaction are usually excluded. The most important agents of metamorphism include temperature, pressure, and fluids. Equally as significant are changes in chemical environment that result in two metamorphic processes: (1) mechanical dislocation where a rock is deformed, especially as a consequence of differential stress; and (2) chemical recrystallization where a mineral assemblage becomes out of equilibrium due to temperature and pressure changes and a new mineral assemblage forms.

Three types of metamorphism may occur depending on the relative effect of mechanical and chemical changes. Dynamic metamorphism, or cataclasis, results mainly from mechanical deformation with little long-term temperature change. Textures produced by such adjustments range from breccias composed of angular, shattered rock fragments to very fine-grained, granulated or powdered rocks with obvious foliation and lineation. Large, pre-existing mineral grains may be deformed as a result of stress. Contact metamorphism occurs primarily as a consequence of increases in temperature when differential stress is minor. A common phenomenon is the effect produced adjacent to igneous intrusions where several metamorphic zones represented by changing mineral assemblages reflect the temperature gradient from the high-temperature intrusion to the low-temperature host rocks; these zones are concentric to the intrusion. Because the volume affected is small, the pressure is near constant. Resulting rocks have equidimensional grains because of a lack of stress and are usually fine-grained due to the short duration of metamorphism. Regional metamorphism results from the general increase, usually correlated, of temperature and pressure over a large area. Grades or intensities of metamorphism are represented by different mineral assemblages that either give relative values of temperature or absolute values when calibrated against laboratory experiments. Regional metamorphism can be subdivided into different pressure-temperature conditions based on observed sequences of mineral assemblages. It may include an extreme condition, where partial melting occurs, called anatexis.

Other types of metamorphism can occur. They are retrograde metamorphism, the response of mineral assemblages to decreasing temperature and pressure; metasomatism, the metamorphism that includes the addition or subtraction of components from the original assemblage; poly-metamorphism, the effect of more than one metamorphic event; and hydrothermal metamorphism, the changes that occur in the presence of water at high temperature and pressure which affect the resulting mineralogy and rate of reaction.

Prograde and retrograde metamorphism

Metamorphism is further divided into prograde and retrograde metamorphism. Prograde metamorphism involves the change of mineral assemblages (paragenesis) with increasing temperature and (usually) pressure conditions. These are solid state dehydration reactions, and involve the loss of volatiles such as water or carbon dioxide. Prograde metamorphism results in rock characteristic of the maximum pressure and temperature experienced. Metamorphic rocks usually do not undergo further change when they are brought back to the surface.

Shock metamorphism or impact metamorphism describes the effects of shock-wave related deformation and heating during impact events. The formation of similar features during explosive volcanism is generally discounted due to the lack of metamorphic effects unequivocally associated with explosions and the difficulty in reaching sufficient pressures during such an event.

EFFECTS

Mineral microstructures

Planar fractures

Planar fractures are parallel sets of multiple planar cracks or cleavages in quartz grains; they develop at the lowest pressures characteristic of shock waves (~5–8 GPa) and a common feature of quartz grains found associated with impact structures. Although the occurrence of planar fractures is relatively common in other deformed rocks, the development of intense, widespread, and closely spaced planar fractures is considered diagnostic of shock metamorphism.

Planar deformation features

Planar deformation features, or PDFs, are optically recognizable microscopic features in grains of silicate minerals (usually quartz or feldspar), consisting of very narrow planes of glassy material arranged in parallel sets that have distinct orientations with respect to the grain's crystal structure. PDFs are only produced by extreme shock compressions on the scale of meteor impacts. They are not found in volcanic environments.

Brazil twinning in quartz

This form of twinning in quartz is relatively common but the occurrence of close-spaced Brazil twins parallel to the basal plane, (0001), has only been reported from impact structures. Experimental formation of basal-orientated Brazil twins in quartz requires high stresses (about 8 GPa) and high strain rates, and it seems probable that such features in natural quartz can also be regarded as unique impact indicators.

High-pressure polymorphs

The very high pressures associated with impacts can lead to the formation of high-pressure polymorphs of various minerals. Quartz may occur as either of its two high-pressure forms, coesite and stishovite. Coesite occasionally occurs associated with eclogites formed during very high pressure regional metamorphism but was first discovered in a meteorite crater in 1960. Stishovite, however, is only known from impact structures. Two of the high-pressure polymorphs of titanium dioxide, one with a baddeleyite-like form and the other with a á-PbO2 structure, have been found associated with the Nördlinger Ries impact structure. Diamond, the high-pressure allotrope of carbon, has been found associated with many impact structures, and both fullerenes and carbynes have been reported.

Shatter cones

Shatter cones have a distinctively conical shape that radiates from the top of the cones repeating cone-on-cone, at various scales in the same sample. They are only known to form in rocks beneath meteorite impact craters or underground nuclear explosions. They are evidence that the rock has been subjected to a shock with pressures in the range of 2-30 GPa.

Occurrence

The effects described above have been found singly, or more often in combination, associated with every impact structure that has been identified on Earth. The search for such effects therefore forms the basis for identifying possible candidate impact structures, particularly to distinguish them from volcanic features.

Metamorphic rock

Metamorphic rocks arise from the transformation of existing rock types, in a process called metamorphism, which means "change in form". The original rock (protolith) is subjected to heat (temperatures greater than 150 to 200 °C) and pressure (1500 bars), causing profound physical and/or chemical change. The protolith may be a sedimentary rock, an igneous rock or another older metamorphic rock.

Metamorphic rocks make up a large part of the Earth's crust and are classified by texture and by chemical and mineral assemblage (metamorphic facies). They may be formed simply by being deep beneath the Earth's surface, subjected to high temperatures and the great pressure of the rock layers above it. They can form from tectonic processes such as continental collisions, which cause horizontal pressure, friction and distortion. They are also formed when rock is heated up by the intrusion of hot molten rock called magma from the Earth's interior. The study of metamorphic rocks (now exposed at the Earth's surface following erosion and uplift) provides information about the temperatures and pressures that occur at great depths within the Earth's crust. Some examples of metamorphic rocks are gneiss, slate, marble, schist, and quartzite.

Metamorphic minerals

Metamorphic minerals are those that form only at the high temperatures and pressures associated with the process of metamorphism. These minerals, known as index minerals, include sillimanite, kyanite, staurolite, andalusite, and some garnet.

Other minerals, such as olivines, pyroxenes, amphiboles, micas, feldspars, and quartz, may be found in metamorphic rocks, but are not necessarily the result of the process of metamorphism. These minerals formed during the crystallization of igneous rocks. They are stable at high temperatures and pressures and may remain chemically unchanged during the metamorphic process. However, all minerals are stable only within certain limits, and the presence of some minerals in metamorphic rocks indicates the approximate temperatures and pressures at which they formed.

The change in the particle size of the rock during the process of metamorphism is called recrystallization. For instance, the small calcite crystals in the sedimentary rock limestone and chalk change into larger crystals in the metamorphic rock marble, or in metamorphosed sandstone, recrystallization of the original quartz sand grains results in very compact quartzite, also known as metaquartzite, in which the often larger quartz crystals are interlocked. Both high temperatures and pressures contribute to recrystallization. High temperatures allow the atoms and ions in solid crystals to migrate, thus reorganizing the crystals, while high pressures cause solution of the crystals within the rock at their point of contact.

Foliation

The layering within metamorphic rocks is called foliation (derived from the Latin word folia, meaning "leaves"), and it occurs when a rock is being shortened along one axis during recrystalli-zation. This causes the platy or elongated crystals of minerals, such as mica and chlorite, to become rotated such that their long axes are perpendicular to the orientation of shortening. This results in a banded, or foliated rock, with the bands showing the colors of the minerals that formed them.

Textures are separated into foliated and non-foliated categories. Foliated rock is a product of differential stress that deforms the rock in one plane, sometimes creating a plane of cleavage. For example, slate is a foliated metamorphic rock, originating from shale. Non-foliated rock does not have planar patterns of strain. Rocks that were subjected to uniform pressure from all sides, or those that lack minerals with distinctive growth habits, will not be foliated. Where a rock has been subject to differential stress, the type of foliation that develops depends on the metamorphic grade. For instance, starting with a mudstone, the following sequence develops with increasing temperature: slate is a very fine-grained, foliated metamorphic rock, characteristic of very low grade metamorphism, while phyllite is fine-grained and found in areas of low grade metamorphism, schist is medium to coarse-grained and found in areas of medium grade metamorphism, and gneiss coarse to very coarse-grained, found in areas of high-grade metamorphism. Marble is generally not foliated, which allows its use as a material for sculpture and architecture.

Another important mechanism of metamorphism is that of chemical reactions that occur between minerals without them melting. In the process atoms are exchanged between the minerals, and thus new minerals are formed. Many complex high-temperature reactions may take place, and each mineral assemblage produced provides us with a clue as to the temperatures and pressures at the time of metamorphism.

Metasomatism is the drastic change in the bulk chemical composition of a rock that often occurs during the processes of metamorphism. It is due to the introduction of chemicals from other surrounding rocks. Water may transport these chemicals rapidly over great distances. Because of the role played by water, metamorphic rocks generally contain many elements absent from the original rock, and lack some that originally were present. Still, the introduction of new chemicals is not necessary for recrystallization to occur.

TYPES OF METAMORPHISM

Contact metamorphism

Contact metamorphism is the name given to the changes that take place when magma is injected into the surrounding solid rock (country rock). The changes that occur are greatest wherever the magma comes into contact with the rock because the temperatures are highest at this boundary and decrease with distance from it. Around the igneous rock that forms from the cooling magma is a metamorphosed zone called a contact metamorphism aureole. Aureoles may show all degrees of metamorphism from the contact area to unmetamorphosed (unchanged) country rock some distance away. The formation of important ore minerals may occur by the process of metasomatism at or near the contact zone.

When a rock is contact altered by an igneous intrusion it very frequently becomes more indurated, and more coarsely crystalline. Many altered rocks of this type were formerly called hornstones, and the term hornfels is often used by geologists to signify those fine grained, compact, non-foliated products of contact metamorphism. A shale may become a dark argillaceous hornfels, full of tiny plates of brownish biotite; a marl or impure limestone may change to a grey, yellow or greenish lime-silicate-hornfels or siliceous marble, tough and splintery, with abundant augite, garnet, wollastonite and other minerals in which calcite is an important component. A diabase or andesite may become a diabase hornfels or andesite hornfels with development of new hornblende and biotite and a partial recrystallization of the original feldspar. Chert or flint may become a finely crystalline quartz rock; sandstones lose their clastic structure and are converted into a mosaic of small close-fitting grains of quartz in a metamorphic rock called quartzite.

If the rock was originally banded or foliated (as, for example, a laminated sandstone or a foliated calc-schist) this character may not be obliterated, and a banded hornfels is the product; fossils even may have their shapes preserved, though entirely recrystallized, and in many contact-altered lavas the vesicles are still visible, though their contents have usually entered into new combinations to form minerals that were not originally present. The minute structures, however, disappear, often completely, if the thermal alteration is very profound. Thus small grains of quartz in a shale are lost or blend with the surrounding particles of clay, and the fine ground-mass of lavas is entirely reconstructed.

By recrystallization in this manner peculiar rocks of very distinct types are often produced. Thus shales may pass into cordierite rocks, or may show large crystals of andalusite (and chiastolite), staurolite, garnet, kyanite and sillimanite, all derived from the aluminous content of the original shale. A considerable amount of mica (both muscovite and biotite) is often simultaneously formed, and the resulting product has a close resemblance to many kinds of schist. Limestones, if pure, are often turned into coarsely crystalline marbles; but if there was an admixture of clay or sand in the original rock such minerals as garnet, epidote, idocrase, wollastonite, will be present. Sandstones when greatly heated may change into coarse quartzites composed of large clear grains of quartz. These more intense stages of alteration are not so commonly seen in igneous rocks, because their minerals, being formed at high temperatures, are not so easily transformed or recrystallized.

In a few cases rocks are fused and in the dark glassy product minute crystals of spinel, sillimanite and cordierite may separate out. Shales are occasionally thus altered by basalt dikes, and feldspathic sandstones may be completely vitrified. Similar changes may be induced in shales by the burning of coal seams or even by an ordinary furnace.

There is also a tendency for metasomatism between the igneous magma and sedimentary country rock, whereby the chemicals in each are exchanged or introduced into the other. Granites may absorb fragments of shale or pieces of basalt. In that case, hybrid rocks called skarn arise, which don't have the characteristics of normal igneous or sedimentary rocks. Sometimes an invading granite magma permeates the rocks around, filling their joints and planes of bedding, etc., with threads of quartz and feldspar. This is very exceptional but instances of it are known and it may take place on a large scale.

Regional metamorphism

Regional metamorphism, also known as dynamic metamorphism, is the name given to changes in great masses of rock over a wide area. Rocks can be metamorphosed simply by being at great depths below the Earth's surface, subjected to high temperatures and the great pressure caused by the immense weight of the rock layers above. Much of the lower continental crust is metamorphic, except for recent igneous intrusions. Horizontal tectonic movements such as the collision of continents create orogenic belts, and cause high temperatures, pressures and deformation in the rocks along these belts. If the metamorphosed rocks are later uplifted and exposed by erosion, they may occur in long belts or other large areas at the surface. The process of metamorphism may have destroyed the original features that could have revealed the rock's previous history. Recrystallization of the rock will destroy the textures and fossils present in sedimentary rocks. Metasomatism will change the original composition.

Regional metamorphism tends to make the rock more indurated and at the same time to give it a foliated, shistose or gneissic texture, consisting of a planar arrangement of the minerals, so that platy or prismatic minerals like mica and hornblende have their longest axes arranged parallel to one another. For that reason many of these rocks split readily in one direction along mica-bearing zones (schists). In gneisses, minerals also tend to be segregated into bands; thus there are seams of quartz and of mica in a mica schist, very thin, but consisting essentially of one mineral.

Along the mineral layers composed of soft or fissile minerals the rocks will split most readily, and the freshly split specimens will appear to be faced or coated with this mineral; for example, a piece of mica schist looked at facewise might be supposed to consist entirely of shining scales of mica. On the edge of the specimens, however, the white folia of granular quartz will be visible. In gneisses these alternating folia are sometimes thicker and less regular than in schists, but most importantly less micaceous; they may be lenticular, dying out rapidly. Gneisses also, as a rule, contain more feldspar than schists do, and are tougher and less fissile. Contortion or crumbling of the foliation is by no means uncommon; splitting faces are undulose or puckered. Schistosity and gneissic banding (the two main types of foliation) are formed by directed pressure at elevated temperature, and to interstitial movement, or internal flow arranging the mineral particles while they are crystallizing in that directed pressure field.

Rocks that were originally sedimentary and rocks that were undoubtedly igneous may be metamorphosed into schists and gneisses. If originally of similar composition they may be very difficult to distinguish from one another if the metamorphism has been great. A quartz-porphyry, for example, and a fine feldspathic sandstone, may both be metamorphosed into a grey or pink mica-schist.

SEDIMENTARY ROCKS AND STRATIGRAPHY 23

Sedimentary rocks are types of rock that are formed by the deposition of material at the Earth's surface and within bodies of water. Sedimentation is the collective name for processes that cause mineral and/or organic particles (detritus) to settle and accumulate or minerals to precipitate from a solution. Particles that form a sedimentary rock by accumulating are called sediment. Before being deposited, sediment was formed by weathering and erosion in a source area, and then transported to the place of deposition by water, wind, ice, mass movement or glaciers which are called agents of denudation.

The sedimentary rock cover of the continents of the Earth's crust is extensive, but the total contribution of sedimentary rocks is estimated to be only 8% of the total volume of the crust. Sedimentary rocks are only a thin veneer over a crust consisting mainly of igneous and metamorphic rocks. Sedimentary rocks are deposited in layers as strata, forming a structure called bedding. The study of sedimentary rocks and rock strata provides information about the subsurface that is useful for civil engineering, for example in the construction of roads, houses, tunnels, canals or other structures. Sedimentary rocks are also important sources of natural resources like coal, fossil fuels, drinking water or ores.

The study of the sequence of sedimentary rock strata is the main source for scientific knowledge about the Earth's history, including palaeogeography, paleoclimatology and the history of life. The scientific discipline that studies the properties and origin of sedimentary rocks is called sedimentology. Sedimentology is part of both geology and physical geography and overlaps partly with other disciplines in the Earth sciences, such as pedology, geomorphology, geochemistry and structural geology.

GENETIC CLASSIFICATION

Based on the processes responsible for their formation, sedimentary rocks can be subdivided into four groups: clastic sedimentary rocks, biochemical (or biogenic) sedimentary rocks, chemical sedimentary rocks and a fourth category for "other" sedimentary rocks formed by impacts, volcanism, and other minor processes.

Clastic sedimentary rocks

Clastic sedimentary rocks are composed of silicate minerals and rock fragments that were transported by moving fluids (as bed load, suspended load, or by sediment gravity flows) and were deposited when these fluids came to rest. Clastic rocks are composed largely of quartz, feldspar, rock (lithic) fragments, clay minerals, and mica; numerous other minerals may be present as accessories and may be important locally.

Clastic sediment, and thus clastic sedimentary rocks, are subdivided according to the dominant particle size (diameter). Most geologists use the Udden-Wentworth grain size scale and divide unconsolidated sediment into three fractions: gravel (>2 mm diameter), sand (1/16 to 2 mm diameter), and mud (clay is <1/256 mm and silt is between 1/16 and 1/256 mm). The classification of clastic sedimentary rocks parallels this scheme; conglomerates and breccias are made mostly of gravel, sandstones are made mostly of sand, and mudrocks are made mostly of mud. This tripartite subdivision is mirrored by the broad categories of rudites, arenites, and lutites, respectively, in older literature.

Subdivision of these three broad categories is based on differences in clast shape (conglomerates and breccias), composition (sandstones), grain size and/or texture (mudrocks).

Conglomerates and breccias

Conglomerates are dominantly composed of rounded gravel and breccias are composed of dominantly angular gravel.

Sandstones

Sandstone classification schemes vary widely, but most geologists have adopted the Dott scheme, which uses the relative abundance of quartz, feldspar, and lithic framework grains and the abundance of muddy matrix between these larger grains.

Composition of framework grains

The relative abundance of sand-sized framework grains determines the first word in a sandstone name. For naming purposes, the abundance of framework grains is normalized to quartz, feldspar, and lithic fragments formed from other rocks. These are the three most abundant components of sandstones; all other minerals are considered accessories and not used in the naming of the rock, regardless of abundance.

- Quartz sandstones have >90% quartz grains

- Feldspathic sandstones have <90% quartz grains and more feldspar grains than lithic grains

- Lithic sandstones have <90% quartz grains and more lithic grains than feldspar grains

Abundance of muddy matrix between sand grains

When sand-sized particles are deposited, the space between the sand grains either remains open or is filled with mud (silt and/or clay sized particle).

- "Clean" sandstones with open pore space (that may later be filled with cement) are called arenites

- Muddy sandstones with abundant (>10%) muddy matrix are called wackes.

Six sandstone names are possible using descriptors for grain composition (quartz-, feldspathic-, and lithic-) and amount of matrix (wacke or arenite). For example, a quartz arenite would be composed of mostly (>90%) quartz grains and have little/no clayey matrix between the grains, a lithic wacke would have abundant lithic grains (<90% quartz, remainder would have more lithics than feldspar) and abundant muddy matrix, etc.

Although the Dott classification scheme is widely used by sedimentologists, common names like greywacke, arkose, and quartz sandstone are still widely used by nonspecialists and in popular literature.

Mudrocks

Mudrocks are sedimentary rocks composed of at least 50% silt- and clay-sized particles. These relatively fine-grained particles are commonly transported as suspended particles by turbulent flow in water or air, and deposited as the flow calms and the particles settle out of suspension.

Most authors presently use the term "mudrock" to refer to all rocks composed dominantly of mud. Mudrocks can be divided into siltstones (composed dominantly of silt-sized particles), mudstones (subequal mixture of silt- and clay-sized particles), and claystones (composed mostly of clay-sized particles). Most authors use "shale" as a term for a fissile mudrock (regardless of grain size) although some older literature uses the term "shale" as a synonym for mudrock.

Biochemical sedimentary rocks

Biochemical sedimentary rocks are created when organisms use materials dissolved in air or water to build their tissue. Examples include:

- Most types of limestone are formed from the calcareous skeletons of organisms such as corals, mollusks, and foraminifera.

- Coal which forms as plants remove carbon from the atmosphere and combine with other elements to build their tissue.

- Deposits of chert formed from the accumulation of siliceous skeletons from microscopic organisms such as radiolaria and diatoms.

Chemical sedimentary rocks

Chemical sedimentary rock forms when mineral constituents in solution become supersaturated and inorganically precipitate. Common chemical sedimentary rocks include oolitic limestone and rocks composed of evaporite minerals such as halite (rock salt), sylvite, barite and gypsum.

"Other" sedimentary rocks

This fourth miscellaneous category includes rocks formed by Pyroclastic flows, impact breccias, volcanic breccias, and other relatively uncommon processes.

Compositional classification schemes

Alternatively, sedimentary rocks can be subdivided into compositional groups based on their mineralogy:

- Siliciclastic sedimentary rocks, as described above, are dominantly composed of silicate minerals. The sediment that makes up these rocks was transported as bed load, suspended load, or by sediment gravity flows. Siliciclastic sedimentary rocks are subdivided into conglomerates and breccias, sandstone, and mudrocks.

- Carbonate sedimentary rocks are composed of calcite (rhombohedral $CaCO_3$), aragonite (orthorhombic $CaCO_3$), dolomite ($CaMg(CO_3)_2$), and other carbonate minerals based on the CO_2" 3 ion. Common examples include limestone and dolostone.

- Evaporite sedimentary rocks are composed of minerals formed from the evaporation of water. The most common evaporite minerals are carbonates (calcite and others based on CO_2"3), chlorides (halite and others built on Cl"), and sulfates (gypsum and others built on SO_2"4). Evaporite rocks commonly include abundant halite (rock salt), gypsum, and anhydrite.

- Organic-rich sedimentary rocks have significant amounts of organic material, generally in excess of 3% total organic carbon. Common examples include coal, oil shale as well as source rocks for oil and natural gas.

- Siliceous sedimentary rocks are almost entirely composed of silica (SiO_2), typically as chert, opal, chalcedony or other microcrystalline forms.

- Iron-rich sedimentary rocks are composed of >15% iron; the most common forms are banded iron formations and ironstones

⊙ Phosphatic sedimentary rocks are composed of phosphate minerals and contain more than 6.5% phosphorus; examples include deposits of phosphate nodules, bone beds, and phosphatic mudrocks

DEPOSITION AND DIAGENESIS

Sediment transport and deposition

Sedimentary rocks are formed when sediment is deposited out of air, ice, wind, gravity, or water flows carrying the particles in suspension. This sediment is often formed when weathering and erosion break down a rock into loose material in a source area. The material is then transported from the source area to the deposition area. The type of sediment transported depends on the geology of the hinterland (the source area of the sediment). However, some sedimentary rocks, like evaporites, are composed of material that formed at the place of deposition. The nature of a sedimentary rock, therefore, not only depends on sediment supply, but also on the sedimentary depositional environment in which it formed.

Diagenesis

The term diagenesis is used to describe all the chemical, physical, and biological changes, including cementation, undergone by a sediment after its initial deposition, exclusive of surface weathering. Some of these processes cause the sediment to consolidate: a compact, solid substance forms out of loose material. Young sedimentary rocks, especially those of Quaternary age (the most recent period of the geologic time scale) are often still unconsolidated. As sediment deposition builds up, the overburden (or lithostatic) pressure rises, and a process known as lithification takes place. Sedimentary rocks are often saturated with seawater or groundwater, in which minerals can dissolve or from which minerals can precipitate. Precipitating minerals reduce the pore space in a rock, a process called cementation. Due to the decrease in pore space, the original connate fluids are expelled. The precipitated minerals form a cement and make the rock more compact and competent. In this way, loose clasts in a sedimentary rock can become "glued" together.

When sedimentation continues, an older rock layer becomes buried deeper as a result. The lithostatic pressure in the rock increases due to the weight of the overlying sediment. This causes compaction, a process in which grains mechanically reorganize. Compaction is, for example, an important diagenetic process in clay, which can initially consist of 60% water. During compaction, this interstitial water is pressed out of pore spaces. Compaction can also be the result of dissolution of grains by pressure solution. The dissolved material precipitates again in open pore spaces, which means there is a net flow of material into the pores. However, in some cases a certain mineral dissolves and does not precipitate again. This process, called leaching, increases pore space in the rock.

Some biochemical processes, like the activity of bacteria, can affect minerals in a rock and are therefore seen as part of diagenesis. Fungi and plants (by their roots) and various other organisms that live beneath the surface can also influence diagenesis.

Burial of rocks due to ongoing sedimentation leads to increased pressure and temperature, which stimulates certain chemical reactions. An example is the reactions by which organic material becomes lignite or coal. When temperature and pressure increase still further, the realm of diagenesis makes way for metamorphism, the process that forms metamorphic rock.

Properties

Color

The color of a sedimentary rock is often mostly determined by iron, an element with two major oxides: iron(II) oxide and iron(III) oxide. Iron(II) oxide only forms under anoxic circumstances and gives the rock a grey or greenish colour. Iron(III) oxide is often in the form of the mineral hematite and gives the rock a reddish to brownish colour. In arid continental climates rocks are in direct contact with the atmosphere, and oxidation is an important process, giving the rock a red or orange colour. Thick sequences of red sedimentary rocks formed in arid climates are called red beds. However, a red colour does not necessarily mean the rock formed in a continental environment or arid climate.

The presence of organic material can colour a rock black or grey. Organic material is in nature formed from dead organisms, mostly plants. Normally, such material eventually decays by oxidation or bacterial activity. Under anoxic circumstances, however, organic material cannot decay and becomes a dark sediment, rich in organic material. This can, for example, occur at the bottom of deep seas and lakes. There is little water current in such environments, so oxygen from surface water is not brought down, and the deposited sediment is normally a fine dark clay. Dark rocks rich in organic material are therefore often shales.

Texture

The size, form and orientation of clasts or minerals in a rock is called its texture. The texture is a small-scale property of a rock, but determined many of its large-scale properties, such as the density, porosity or permeability.

Clastic rocks have a 'clastic texture', which means they consist of clasts. The 3D orientation of these clasts is called the fabric of the rock. Between the clasts, the rock can be composed of a matrix or a cement (the latter can consist of crystals of one or more precipitated minerals). The size and form of clasts can be used to determine the velocity and direction of current in the sedimentary environment where the rock was formed; fine, calcareous mud only settles in quiet water while gravel and larger clasts are only deposited by rapidly moving water. The grain size of a rock is usually expressed with the Wentworth scale, though alternative scales are sometimes used. The grain size can be expressed as a diameter or a volume, and is always an average value – a rock is composed of clasts with different sizes. The statistical distribution of grain sizes is different for different rock types and is described in a property called the sorting of the rock. When all clasts are more or less of the same size, the rock is called 'well-sorted', and when there is a large spread in grain size, the rock is called 'poorly sorted'.

The form of clasts can reflect the origin of the rock.

Coquina, a rock composed of clasts of broken shells, can only form in energetic water. The form of a clast can be described by using four parameters:

◉ Surface texture describes the amount of small-scale relief of the surface of a grain that is too small to influence the general shape.

◉ Rounding describes the general smoothness of the shape of a grain.

◉ 'Sphericity' describes the degree to which the grain approaches a sphere.

◉ 'Grain form' describes the three dimensional shape of the grain.

Chemical sedimentary rocks have a non-clastic texture, consisting entirely of crystals. To describe such a texture, only the average size of the crystals and the fabric are necessary.

Mineralogy

Most sedimentary rocks contain either quartz (especially siliciclastic rocks) or calcite (especially carbonate rocks). In contrast to igneous and metamorphic rocks, a sedimentary rock usually contains very few different major minerals. However, the origin of the minerals in a sedimentary rock is often more complex than in an igneous rock. Minerals in a sedimentary rock can have formed by precipitation during sedimentation or by diagenesis. In the second case, the mineral precipitate can have grown over an older generation of cement. A complex diagenetic history can be studied by optical mineralogy, using a petrographic microscope.

Carbonate rocks dominantly consist of carbonate minerals like calcite, aragonite or dolomite. Both cement and clasts (including fossils and ooids) of a carbonate rock can consist of carbonate minerals. The mineralogy of a clastic rock is determined by the supplied material from the source area, the manner of transport to the place of deposition and the stability of a particular mineral. The stability of the major rock forming minerals (their resistance to weathering) is expressed by Bowen's reaction series. In this series, quartz is the most stable, followed by feldspar, micas, and other less stable minerals that are only present when little weathering has occurred. The amount of weathering depends mainly on the distance to the source area, the local climate and the time it took for the sediment to be transported there. In most sedimentary rocks, mica, feldspar and less stable minerals have reacted to clay minerals like kaolinite, illite or smectite.

Fossils

Among the three major types of rock, fossils are most commonly found in sedimentary rock. Unlike most igneous and metamorphic rocks, sedimentary rocks form at temperatures and pressures that do not destroy fossil remnants. Often these fossils may only be visible when studied under a microscope (microfossils) or with a loupe.

Dead organisms in nature are usually quickly removed by scavengers, bacteria, rotting and erosion, but sedimentation can contribute to exceptional circumstances where these natural processes are unable to work, causing fossilisation. The chance of fossilisation is higher when the sedimentation rate is high (so that a carcass is quickly buried), in anoxic environments (where little bacterial activity occurs) or when the organism had a particularly hard skeleton. Larger, well-preserved fossils are relatively rare.

Fossils can be both the direct remains or imprints of organisms and their skeletons. Most commonly preserved are the harder parts of organisms such as bones, shells, and the woody tissue of plants. Soft tissue has a much smaller chance of being preserved and fossilized, and soft tissue of animals older than 40 million years is very rare. Imprints of organisms made while still alive are called trace fossils. Examples are burrows, footprints, etc.

Being part of a sedimentary or metamorphic rock, fossils undergo the same diagenetic processes as rock. A shell consisting of calcite can, for example, dissolve while a cement of silica then fills the cavity. In the same way, precipitating minerals can fill cavities formerly occupied by blood vessels, vascular tissue or other soft tissues. This preserves the form of the organism but changes the chemical composition, a process called permineralization. The most common minerals in permineralization cements are carbonates (especially calcite), forms of amorphous silica (chalcedony, flint, chert) and pyrite. In the case of silica cements, the process is called lithification.

At high pressure and temperature, the organic material of a dead organism undergoes chemical reactions in which volatiles like water and carbon dioxide are expelled. The fossil, in the end, consists of a thin layer of pure carbon or its mineralized form, graphite. This form of fossilisation is called carbonisation. It is particularly important for plant fossils. The same process is responsible for the formation of fossil fuels like lignite or coal.

Primary sedimentary structures

Structures in sedimentary rocks can be divided into 'primary' structures (formed during deposition) and 'secondary' structures (formed after deposition). Unlike textures, structures are always large-scale features that can easily be studied in the field. Sedimentary structures can indicate something about the sedimentary environment or can serve to tell which side originally faced up where tectonics have tilted or overturned sedimentary layers.

Sedimentary rocks are laid down in layers called beds or strata. A bed is defined as a layer of rock that has a uniform lithology and texture. Beds form by the deposition of layers of sediment on top of each other. The sequence of beds that characterizes sedimentary rocks is called bedding. Single beds can be a couple of centimetres to several meters thick. Finer, less pronounced layers are called laminae, and the structure it forms in a rock is called lamination. Laminae are usually less than a few centimetres thick. Though bedding and lamination are often originally horizontal in nature, this is not always the case. In some environments, beds are deposited at a (usually small) angle. Sometimes multiple sets of layers with different orientations exist in the same rock, a structure called cross-bedding. Cross-bedding forms when small-scale erosion occurs during deposition, cutting off part of the beds. Newer beds then form at an angle to older ones.

The opposite of cross-bedding is parallel lamination, where all sedimentary layering is parallel. With laminations, differences are generally caused by cyclic changes in the sediment supply, caused, for example, by seasonal changes in rainfall, temperature or biochemical activity. Laminae that represent seasonal changes (similar to tree rings) are called varves. Any sedimentary rock composed of millimeter or finer scale layers can be named with the general term laminite. Some rocks have no lamination at all; their structural character is called massive bedding.

Graded bedding is a structure where beds with a smaller grain size occur on top of beds with larger grains. This structure forms when fast flowing water stops flowing. Larger, heavier clasts in suspension settle first, then smaller clasts. Though graded bedding can form in many different environments, it is characteristic for turbidity currents.

The bedform (the surface of a particular bed) can be indicative for a particular sedimentary environment, too. Examples of bed forms include dunes and ripple marks. Sole markings, such as tool marks and flute casts, are groves dug into a sedimentary layer that are preserved. These are often elongated structures and can be used to establish the direction of the flow during deposition.

Ripple marks also form in flowing water. There are two types: asymmetric wave ripples and symmetric current ripples. Environments where the current is in one direction, such as rivers, produce asymmetric ripples. The longer flank of such ripples is oriented opposite to the direction of the current. Wave ripples occur in environments where currents occur in all directions, such as tidal flats. Mudcracks are a bed form caused by the dehydration of sediment that occasionally comes above the water surface. Such structures are commonly found at tidal flats or point bars along rivers.

SECONDARY SEDIMENTARY STRUCTURES

Secondary sedimentary structures are structures in sedimentary rocks which formed after deposition. Such structures form by chemical, physical and biological processes inside the sediment. They can be indicators for circumstances after deposition. Some can be used as way up criteria.

Organic presence in a sediment can leave more traces than just fossils. Preserved tracks and burrows are examples of trace fossils (also called ichnofossils). Some trace fossils such as paw prints of dinosaurs or early humans can capture human imagination, but such traces are relatively rare. Most trace fossils are burrows of molluscs or arthropods. This burrowing is called bioturbation by sedimentologists. It can be a valuable indicator of the biological and ecological environment after the sediment was deposited. On the other hand, the burrowing activity of organisms can destroy other (primary) structures in the sediment, making a reconstruction more difficult.

Secondary structures can also have been formed by diagenesis or the formation of a soil (pedogenesis) when a sediment is exposed above the water level. An example of a diagenetic structure common in carbonate rocks is a stylolite. Stylolites are irregular planes where material was dissolved into the pore fluids in the rock. The result of precipitation of a certain chemical species can be colouring and staining of the rock, or the formation of concretions. Concretions are roughly concentric bodies with a different composition from the host rock. Their formation can be the result of localized precipitation due to small differences in composition or porosity of the host rock, such as around fossils, inside burrows or around plant roots. In carbonate rocks such as limestone or chalk, chert or flint concretions are common, while terrestrial sandstones can have iron concretions. Calcite concretions in clay are called septarian concretions.

After deposition, physical processes can deform the sediment, forming a third class of secondary structures. Density contrasts between different sedimentary layers, such as between sand and clay, can result in flame structures or load casts, formed by inverted diapirism. The diapirism causes the denser upper layer to sink into the other layer. Sometimes, density contrast can result or grow when one of the lithologies dehydrates. Clay can be easily compressed as a result of dehydration, while sand retains the same volume and becomes relatively less dense. On the other hand, when the pore fluid pressure in a sand layer surpasses a critical point, the sand can flow through overlying clay layers, forming discordant bodies of sedimentary rock called sedimentary dykes (the same process can form mud volcanoes on the surface).

A sedimentary dyke can also be formed in a cold climate where the soil is permanently frozen during a large part of the year. Frost weathering can form cracks in the soil that fill with rubble from above. Such structures can be used as climate indicators as well as way up structures.

Density contrasts can also cause small-scale faulting, even while sedimentation goes on (syn-sedimentary faulting). Such faulting can also occur when large masses of non-lithified sediment are deposited on a slope, such as at the front side of a delta or the continental slope. Instabilities in such sediments can result in slumping. The resulting structures in the rock are syn-sedimentary folds and faults, which can be difficult to distinguish from folds and faults formed by tectonic forces in lithified rocks.

Sedimentary environments

The setting in which a sedimentary rock forms is called the sedimentary environment. Every environment has a characteristic combination of geologic processes and circumstances. The type of sediment that is deposited is not only dependent on the sediment that is transported to a place, but also on the environment itself.

A marine environment means the rock was formed in a sea or ocean. Often, a distinction is made between deep and shallow marine environments. Deep marine usually refers to environments more than 200 m below the water surface. Shallow marine environments exist adjacent to coastlines and can extend out to the boundaries of the continental shelf. The water in such environments has a generally higher energy than that in deep environments, because of wave activity. This means coarser sediment particles can be transported and the deposited sediment can be coarser than in deep environments. When the available sediment is transported from the continent, an alternation of sand, clay and silt is deposited. When the continent is far away, the amount of such sediment brought in may be small, and biochemical processes dominate the type of rock that forms. Especially in warm climates, shallow marine environments far offshore mainly see deposition of carbonate rocks. The shallow, warm water is an ideal habitat for many small organisms that build carbonate skeletons. When these organisms die their skeletons sink to the bottom, forming a thick layer of calcareous mud that may lithify into limestone. Warm shallow marine environments also are ideal environments for coral reefs, where the sediment consists mainly of the calcareous skeletons of larger organisms.

In deep marine environments, the water current over the sea bottom is small. Only fine particles can be transported to such places. Typically sediments depositing on the ocean floor are fine clay or small skeletons of micro-organisms. At 4 km depth, the solubility of carbonates increases dramatically (the depth zone where this happens is called the lysocline). Calcareous sediment that sinks below the lysocline dissolve, so no limestone can be formed below this depth. Skeletons of micro-organisms formed of silica (such as radiolarians) still deposit though. An example of a rock formed out of silica skeletons is radiolarite. When the bottom of the sea has a small inclination, for example at the continental slopes, the sedimentary cover can become unstable, causing turbidity currents. Turbidity currents are sudden disturbances of the normally quite deep marine environment and can cause the geologically speaking instantaneous deposition of large amounts of sediment, such as sand and silt. The rock sequence formed by a turbidity current is called a turbidite.

The coast is an environment dominated by wave action. At the beach, dominantly coarse sediment like sand or gravel is deposited, often mingled with shell fragments. Tidal flats and shoals are places that sometimes dry out because of the tide. They are often cross-cut by gullies, where the current is strong and the grain size of the deposited sediment is larger. Where along a coast (either the coast of a sea or a lake) rivers enter the body of water, deltas can form. These are large accumulations of sediment transported from the continent to places in front of the mouth of the river. Deltas are dominantly composed of clastic sediment.

A sedimentary rock formed on the land has a continental sedimentary environment. Examples of continental environments are lagoons, lakes, swamps, floodplains and alluvial fans. In the quiet water of swamps, lakes and lagoons, fine sediment is deposited, mingled with organic material from dead plants and animals. In rivers, the energy of the water is much higher and the transported material consists of clastic sediment. Besides transport by water, sediment can in continental environments also be transported by wind or glaciers. Sediment transported by wind is called aeolian and is always very well sorted, while sediment transported by a glacier is called glacial till and is characterized by very poor sorting.

Aeolian deposits can be quite striking. The depositional environment of the Touchet Formation, located in the Northwestern United States, had intervening periods of aridity which resulted in a series of rhythmite layers. Erosional cracks were later infilled with layers of soil material, especially from aeolian processes. The infilled sections formed vertical inclusions in the horizontally deposited layers of the Touchet Formation, and thus provided evidence of the events that intervened in time among the forty-one layers that were deposited.

Sedimentary facies

Sedimentary environments usually exist alongside each other in certain natural successions. A beach, where sand and gravel is deposited, is usually bounded by a deeper marine environment a little offshore, where finer sediments are deposited at the same time. Behind the beach, there can be dunes (where the dominant deposition is well sorted sand) or a lagoon (where fine clay and organic material is deposited). Every sedimentary environment has its own characteristic deposits. The typical rock formed in a certain environment is called its sedimentary facies. When sedimentary strata accumulate through time, the environment can shift, forming a change in facies in the subsurface at one location. On the other hand, when a rock layer with a certain age is followed laterally, the lithology (the type of rock) and facies eventually change.

Facies can be distinguished in a number of ways: the most common ways are by the lithology (for example: limestone, siltstone or sandstone) or by fossil content. Coral for example only lives in warm and shallow marine environments and fossils of coral are thus typical for shallow marine facies. Facies determined by lithology are called lithofacies; facies determined by fossils are biofacies.

Sedimentary environments can shift their geographical positions through time. Coastlines can shift in the direction of the sea when the sea level drops, when the surface rises due to tectonic forces in the Earth's crust or when a river forms a large delta. In the subsurface, such geographic shifts of sedimentary environments of the past are recorded in shifts in sedimentary facies. This means that sedimentary facies can change either parallel or perpendicular to an imaginary layer of rock with a fixed age, a phenomenon described by Walther's Law.

The situation in which coastlines move in the direction of the continent is called transgression. In the case of transgression, deeper marine facies are deposited over shallower facies, a succession called onlap. Regression is the situation in which a coastline moves in the direction of the sea. With regression, shallower facies are deposited on top of deeper facies, a situation called offlap.

The facies of all rocks of a certain age can be plotted on a map to give an overview of the palaeogeography. A sequence of maps for different ages can give an insight in the development of the regional geography.

Sedimentary basins

Places where large-scale sedimentation takes place are called sedimentary basins. The amount of sediment that can be deposited in a basin depends on the depth of the basin, the so-called accommodation space. Depth, shape and size of a basin depend on tectonics, movements within the Earth's lithosphere. Where the lithosphere moves upward (tectonic uplift), land eventually rises above sea level, so that and erosion removes material, and the area becomes a source for new sediment. Where the lithosphere moves downward (tectonic subsidence), a basin forms and sedimentation can take place. When the lithosphere keeps subsiding, new accommodation space keeps being created.

A type of basin formed by the moving apart of two pieces of a continent is called a rift basin. Rift basins are elongated, narrow and deep basins. Due to divergent movement, the lithosphere is stretched and thinned, so that the hot asthenosphere rises and heats the overlying rift basin. Apart from continental sediments, rift basins normally also have part of their infill consisting of volcanic deposits. When the basin grows due to continued stretching of the lithosphere, the rift grows and the sea can enter, forming marine deposits.

When a piece of lithosphere that was heated and stretched cools again, its density rises, causing isostatic subsidence. If this subsidence continues long enough the basin is called a sag basin. Examples of sag basins are the regions along passive continental margins, but sag basins can also be found in the interior of continents. In sag basins, the extra weight of the newly deposited sediments is enough to keep the subsidence going in a vicious circle. The total thickness of the sedimentary infill in a sag basins can thus exceed 10 km.

A third type of basin exists along convergent plate boundaries - places where one tectonic plate moves under another into the asthenosphere. The subducting plate bends and forms a fore-arc basin in front of the overriding plate—an elongated, deep asymmetric basin. Fore-arc basins are filled with deep marine deposits and thick sequences of turbidites. Such infill is called flysch. When the convergent movement of the two plates results in continental collision, the basin becomes shallower and develops into a foreland basin. At the same time, tectonic uplift forms a mountain belt in the overriding plate, from which large amounts of material are eroded and transported to the basin. Such erosional material of a growing mountain chain is called molasse and has either a shallow marine or a continental facies.

At the same time, the growing weight of the mountain belt can cause isostatic subsidence in the area of the overriding plate on the other side to the mountain belt. The basin type resulting from this subsidence is called a back-arc basin and is usually filled by shallow marine deposits and molasse.

Influence of astronomical cycles

In many cases facies changes and other lithological features in sequences of sedimentary rock have a cyclic nature. This cyclic nature was caused by cyclic changes in sediment supply and the sedimentary environment. Most of these cyclic changes are caused by astronomic cycles. Short astronomic cycles can be the difference between the tides or the spring tide every two weeks. On a larger time-scale, cyclic changes in climate and sea level are caused by Milankovitch cycles: cyclic changes in the orientation and/or position of the Earth's rotational axis and orbit around the Sun. There are a number of Milankovitch cycles known, lasting between 10,000 and 200,000 years.

Relatively small changes in the orientation of the Earth's axis or length of the seasons can be a major influence on the Earth's climate. An example are the ice ages of the past 2.6 million years (the Quaternary period), which are assumed to have been caused by astronomic cycles. Climate change can influence the global sea level (and thus the amount of accommodation space in sedimentary basins) and sediment supply from a certain region. Eventually, small changes in astronomic parameters can cause large changes in sedimentary environment and sedimentation.

Sedimentation rates

The rate at which sediment is deposited differs depending on the location. A channel in a tidal flat can see the deposition of a few metres of sediment in one day, while on the deep ocean floor each year only a few millimetres of sediment accumulate. A distinction can be made between normal sedimentation and sedimentation caused by catastrophic processes. The latter category includes all kinds of sudden exceptional processes like mass movements, rock slides or flooding. Catastrophic processes can see the sudden deposition of a large amount of sediment at once. In some sedimentary environments, most of the total column of sedimentary rock was formed by catastrophic processes, even though the environment is usually a quiet place. Other sedimentary environments are dominated by normal, ongoing sedimentation.

In many cases, sedimentation occurs slowly. In a desert, for example, the wind deposits siliciclastic material (sand or silt) in some spots, or catastrophic flooding of a wadi may cause sudden deposits of large quantities of detrital material, but in most places eolian erosion dominates. The amount of sedimentary rock that forms is not only dependent on the amount of supplied material, but also on how well the material consolidates. Erosion removes most deposited sediment shortly after deposition.

Stratigraphy

That new rock layers are above older rock layers is stated in the principle of superposition. There are usually some gaps in the sequence called unconformities. These represent periods where no new sediments were laid down, or when earlier sedimentary layers raised above sea level and eroded away.

Sedimentary rocks contain important information about the history of the Earth. They contain fossils, the preserved remains of ancient plants and animals. Coal is considered a type of sedimentary rock. The composition of sediments provides us with clues as to the original rock. Differences between successive layers indicate changes to the environment over time. Sedimentary rocks can contain fossils because, unlike most igneous and metamorphic rocks, they form at temperatures and pressures that do not destroy fossil remains.

Stratigraphy

Stratigraphy is a branch of geology which studies rock layers (strata) and layering (stratification). It is primarily used in the study of sedimentary and layered volcanic rocks. Stratigraphy includes two related subfields: lithologic stratigraphy or lithostratigraphy, and biologic stratigraphy or biostratigraphy.

Historical development

Nicholas Steno established the theoretical basis for stratigraphy when he introduced the law of superposition, the principle of original horizontality and the principle of lateral continuity in a 1669 work on the fossilization of organic remains in layers of sediment. The first practical large-scale application of stratigraphy was by William Smith in the 1790s and early 1800s. Smith, known as the "Father of English geology", created the first geologic map of England and first recognized the significance of strata or rock layering and the importance of fossil markers for correlating strata. Another influential application of stratigraphy in the early 1800s was a study by Georges Cuvier and Alexandre Brongniart of the geology of the region around Paris.

Lithostratigraphy

Lithostratigraphy, or lithologic stratigraphy, provides the most obvious visible layering. It deals with the physical contrasts in lithology, or rock type. Such layers can occur both vertically – in layering or bedding of varying rock types – and laterally – reflecting changing environments of deposition (known as facies change). Key concepts in stratigraphy involve understanding how certain geometric relationships between rock layers arise and what these geometries mean in terms of the depositional environment. Stratigraphers have codified a basic concept of their discipline in the law of superposition, which simply states that, in an undeformed stratigraphic sequence, the oldest strata occur at the base of the sequence.

Chemostratigraphy studies the changes in the relative proportions of trace elements and isotopes within and between lithologic units. Carbon and oxygen isotope ratios vary with time, and researchers can use them to map subtle changes that occurred in the paleoenvironment. This has led to the specialized field of isotopic stratigraphy.

Cyclostratigraphy documents the often cyclic changes in the relative proportions of minerals (particularly carbonates), grain size, or thickness of sediment layers (varves) and of fossil diversity with time, related to seasonal or longer term changes in palaeoclimates.

Biostratigraphy

Biostratigraphy or paleontologic stratigraphy is based on fossil evidence in the rock layers. Strata from widespread locations containing the same fossil fauna and flora are correlatable in time. Biologic stratigraphy was based on William Smith's principle of faunal succession, which predated, and was one of the first and most powerful lines of evidence for, biological evolution. It provides strong evidence for the formation (speciation) and extinction of species. The geologic time scale was developed during the 19th century, based on the evidence of biologic stratigraphy and faunal succession. This timescale remained a relative scale until the development of radiometric dating, which gave it and the stratigraphy it was based on an absolute time framework, leading to the development of chronostratigraphy.

One important development is the Vail curve, which attempts to define a global historical sea-level curve according to inferences from worldwide stratigraphic patterns. Stratigraphy is also commonly used to delineate the nature and extent of hydrocarbon-bearing reservoir rocks, seals, and traps in petroleum geology.

Chronostratigraphy

Chronostratigraphy is the branch of stratigraphy that studies the absolute, not relative, age of rock strata. The branch is concerned with deriving geochronological data for rock units, both directly and inferentially, so that a sequence of time-relative events of rocks within a region can be derived. In essence, chronostratigraphy seeks to understand the geologic history of rocks and regions. The ultimate aim of chronostratigraphy is to arrange the sequence of deposition and the time of deposition of all rocks within a geological region and, eventually, the entire geologic record of the earth.

A gap or missing strata in the geological record of an area is called a stratigraphic hiatus. This may be the result of lack of sediment deposition or it may be due to removal by erosion, in which case it may be called a vacuity. It is called a hiatus because deposition was on hold for a period of time. A physical gap may represent both a period of non-deposition and a period of erosion. A fault may cause the appearance of a hiatus.

MAGNETOSTRATIGRAPHY

Magnetostratigraphy is a chronostratigraphic technique used to date sedimentary and volcanic sequences. The method works by collecting oriented samples at measured intervals throughout a section. The samples are analyzed to determine their detrital remanent magnetism (DRM), that is, the polarity of Earth's magnetic field at the time a stratum was deposited. For sedimentary rocks, this is possible because, when very fine-grained magnetic minerals (< 17 ìm) fall through the water column, they orient themselves with Earth's magnetic field. Upon burial, that orientation is preserved. The minerals behave like tiny compasses. For volcanic rocks, magnetic minerals, which form in the melt, are fixed in place upon crystallization or freezing of the lava and are oriented with the ambient magnetic field.

Oriented paleomagnetic core samples are collected in the field; mudstones, siltstones, and very fine-grained sandstones are the preferred lithologies because the magnetic grains are finer and more likely to orient with the ambient field during deposition. If the ancient magnetic field were oriented similar to today's field (North Magnetic Pole near the North Rotational Pole), the strata would retain a normal polarity. If the data indicate that the North Magnetic Pole were near the South Rotational Pole, the strata would exhibit reversed polarity.

Results of the individual samples are analyzed by removing the natural remanent magnetization (NRM) to reveal the DRM. Following statistical analysis, the results are used to generate a local magnetostratigraphic column that can then be compared against the Global Magnetic Polarity Time Scale.

This technique is used to date sequences that generally lack fossils or interbedded igneous rocks. The continuous nature of the sampling means that it is also a powerful technique for the estimation of sediment-accumulation rates.

Archaeological stratigraphy

In the field of archaeology, soil stratigraphy is used to better understand the processes that form and protect archaeological sites. Since the law of superposition holds true, it can help date finds or features from each context; these finds and features can be placed in sequence and the dates interpolated. Phases of activity can also often be seen through stratigraphy, especially when a trench or feature is viewed in section (profile). Because pits and other features can be dug down into earlier levels, not all material at the same absolute depth is necessarily of the same age; close attention has to be paid to the archeological layers. The Harris-matrix is a tool to depict complex stratigraphic relations when they are found, for example, in the context of urban archaeology.

GEOLOGICAL STRUCTURES: FOLD AND FAULT

FOLD (GEOLOGY)

A geological fold occurs when one or a stack of originally flat and planar surfaces, such as sedimentary strata, are bent or curved as a result of permanent deformation. Synsedimentary folds are those due to slumping of sedimentary material before it is lithified. Folds in rocks vary in size from microscopic crinkles to mountain-sized folds. They occur singly as isolated folds and in extensive fold trains of different sizes, on a variety of scales.

Folds form under varied conditions of stress, hydrostatic pressure, pore pressure, and temperature gradient, as evidenced by their presence in soft sediments, the full spectrum of metamorphic rocks, and even as primary flow structures in some igneous rocks. A set of folds distributed on a regional scale constitutes a fold belt, a common feature of orogenic zones. Folds are commonly formed by shortening of existing layers, but may also be formed as a result of displacement on a non-planar fault (fault bend fold), at the tip of a propagating fault (fault propagation fold), by differential compaction or due to the effects of a high-level igneous intrusion e.g. above a laccolith.

Describing folds

Folds are classified by their size, fold shape, tightness, and dip of the axial plane.

Fold terminology in two dimensions

Looking at a fold surface in profile the fold can be divided into hinge and limb portions. The limbs are the flanks of the fold and the hinge is where the flanks join together. The hinge point is the point of minimum radius of curvature (maximum curvature) for a fold. The crest of the fold is the highest point of the fold surface, and the trough is the lowest point. The inflection point of a fold is the point on a limb at which the concavity reverses; on regular folds, this is the midpoint of the limb.

Fold terminology in three dimensions

The hinge points along an entire folded surface form a hinge line, which can be either a crest line or a trough line. The trend and plunge of a linear hinge line gives you information about the orientation of the fold. To more completely describe the orientation of a fold, one must describe the axial surface. The axial surface is the surface defined by connecting all the hinge lines of stacked folding surfaces. If the axial surface is a planar surface then it is called the axial plane and can be described by the strike and dip of the plane. An axial trace is the line of intersection of the axial surface with any other surface (ground, side of mountain, geological cross-section).

Finally, folds can have, but don't necessarily have a fold axis. A fold axis, "is the closest approximation to a straight line that when moved parallel to itself, generates the form of the fold." (Davis and Reynolds, 1996 after Donath and Parker, 1964; Ramsay 1967). A fold that can be generated by a fold axis is called a cylindrical fold. This term has been broadened to include near-cylindrical folds. Often, the fold axis is the same as the hinge line.

Fold shape

A fold can be shaped as a chevron, with planar limbs meeting at an angular axis, as cuspate with curved limbs, as circular with a curved axis, or as elliptical with unequal wavelength.

Fold tightness

Fold tightness is defined by the size of the angle between the fold's limbs (as measured tangential to the folded surface at the inflection line of each limb), called the interlimb angle. Gentle folds have an interlimb angle of between 180° and 120°, open folds range from 120° to 70°, close folds from 70° to 30°, and tight folds from 30° to 0°. Isoclines, or isoclinal folds, have an interlimb angle of between 10° and zero, with essentially parallel limbs.

Fold symmetry

Not all folds are equal on both sides of the axis of the fold. Those with limbs of relatively equal length are termed symmetrical, and those with highly unequal limbs are asymmetrical. Asymme-trical folds generally have an axis at an angle to the original unfolded surface they formed on.

Deformation style classes

Folds that maintain uniform layer thickness are classed as concentric folds. Those that do not are called similar folds. Similar folds tend to display thinning of the limbs and thickening of the hinge zone. Concentric folds are caused by warping from active buckling of the layers, whereas similar folds usually form by some form of shear flow where the layers are not mechanically active. Ramsay has proposed a classification scheme for folds that often is used to describe folds in profile based upon curvature of the inner and outer lines of a fold, and the behavior of dip isogons. that is, lines connecting points of equal dip:

Fold types

- ◉ Anticline: linear, strata normally dip away from axial center, oldest strata in center.

- ◉ Syncline: linear, strata normally dip toward axial center, youngest strata in center.

- ◉ Antiform: linear, strata dip away from axial center, age unknown, or inverted.

- ◉ Synform: linear, strata dip toward axial centre, age unknown, or inverted.

- ◉ Dome: nonlinear, strata dip away from center in all directions, oldest strata in center.

- ◉ Basin: nonlinear, strata dip toward center in all directions, youngest strata in center.

- ◉ Monocline: linear, strata dip in one direction between horizontal layers on each side.

- ◉ Chevron: angular fold with straight limbs and small hinges

- ◉ Recumbent: linear, fold axial plane oriented at low angle resulting in overturned strata in one limb of the fold.

- ◉ Slump: typically monoclinal, result of differential compaction or dissolution during sedimentation and lithification.

- ◉ Ptygmatic: Folds are chaotic, random and disconnected. Typical of sedimentary slump folding, migmatites and decollement detachment zones.

- ◉ Parasitic: short wavelength folds formed within a larger wavelength fold structure-normally associated with differences in bed thickness

(A homocline involves strata dipping in the same direction, though not necessarily any folding.)

Causes of folding

Folds appear on all scales, in all rock types, at all levels in the crust and arise from a variety of causes.

Layer-parallel shortening

When a sequence of layered rocks is shortened parallel to its layering, this deformation may be accommodated in a number of ways, homogeneous shortening, reverse faulting or folding. The response depends on the thickness of the mechanical layering and the contrast in properties between the layers. If the layering does begin to fold, the fold style is also dependent on these properties. Isolated thick competent layers in a less competent matrix control the folding and typically generate classic rounded buckle folds accommodated by deformation in the matrix. In the case of regular alternations of layers of contrasting properties, such as sandstone-shale sequences, kink-bands, box-folds and chevron folds are normally produced.

Fault-related folding

Many folds are directly related to faults, associate with their propagation, displacement and the accommodation of strains between neighbouring faults.

Fault bend folding

Fault bend folds are caused by displacement along a non-planar fault. In non-vertical faults, the hanging-wall deforms to accommodate the mismatch across the fault as displacement progresses. Fault bend folds occur in both extensional and thrust faulting. In extension, listric faults form rollover anticlines in their hanging walls. In thrusting, ramp anticlines are formed whenever a thrust fault cuts up section from one detachment level to another. Displacement over this higher-angle ramp generates the folding.

Fault propagation folding

Fault propagation folds or tip-line folds are caused when displacement occurs on an existing fault without further propagation. In both reverse and normal faults this leads to folding of the overlying sequence, often in the form of a monocline.

Detachment folding

When a thrust fault continues to displace above a planar detachment without further fault propagation, detachment folds may form, typically of box-fold style. These generally occur above a good detachment such as in the Jura Mountains, where the detachment occurs on middle Triassic evaporites.

Folding in shear zones

Shear zones that approximate to simple shear typically contain minor asymmetric folds, with the direction of overturning consistent with the overall shear sense. Some of these folds have highly curved hinge lines and are referred to as sheath folds. Folds in shear zones can be inherited, formed due to the orientation of pre-shearing layering or formed due to instability within the shear flow.

Folding in sediments

Recently deposited sediments are normally mechanically weak and prone to remobilisation before they become lithified, leading to folding. To distinguish them from folds of tectonic origin such structures are called synsedimentary (formed during sedimentation).

Slump folding: When slumps form in poorly consolidated sediments they commonly undergo folding, particularly at their leading edges, during their emplacement. The asymmetry of the slump folds can be used to determine paleoslope directions in sequences of sedimentary rocks.

Dewatering: Rapid dewatering of sandy sediments, possibly triggered by seismic activity can cause convolute bedding.

Compaction: Folds can be generated in a younger sequence by differential compaction over older structures such as fault blocks and reefs.

Igneous intrusion

The emplacement of igneous intrusions tends to deform the surrounding country rock. In the case of high-level intrusions, near the Earth's surface, this deformation is concentrated above the intrusion and often takes the form of folding, as with the upper surface of a laccolith.

Flow folding

The compliance of rock layers is referred to as competence: a competent layer or bed of rock can withstand an applied load without collapsing and is relatively strong, while an incompetent layer is relatively weak. When rock behaves as a fluid, as in the case of very weak rock such as rock salt, or any rock that is buried deeply enough, they typically show flow folding (also called passive folding, because little resistance is offered): the strata appear shifted undistorted, assuming any shape impressed upon them by surrounding more rigid rocks. The strata simply serve as markers of the folding. Such folding is also a feature of many igneous intrusions and glacier ice.

Folding mechanisms

Folding of rocks must balance the deformation of layers with the conservation of volume in a rock mass. This occurs by several mechanisms.

Flexural slip

Flexural slip allows folding by creating layer-parallel slip between the layers of the folded strata, which, altogether, result in deformation. A good analogy is bending a phone book, where volume preservation is accommodated by slip between the pages of the book.

The fold formed by the compression of competent rock beds are called "flexure fold"

Buckling

Typically, folding is thought to occur by simple buckling of a planar surface and its confining volume. The volume change is accommodated by layer parallel shortening the volume, which grows in thickness. Folding under this mechanism is typically of the similar fold style, as thinned limbs are shortened horizontally and thickened hinges do so vertically.

Mass displacement

If the folding deformation cannot be accommodated by flexural slip or volume-change shortening (buckling), the rocks are generally removed from the path of the stress. This is achieved by pressure dissolution, a form of metamorphic process, in which rocks shorten by dissolving constituents in areas of high strain and redepositing them in areas of lower strain. Folds created in this way include examples in migmatites, and areas with a strong axial planar cleavage.

Mechanics of folding

Folds in rock are formed in relation to the stress field in which the rocks are located and the rheology, or method of response to stress, of the rock at the time at which the stress is applied.

The rheology of the layers being folded determines characteristic features of the folds that are measured in the field. Rocks that deform more easily form many short-wavelength, high-amplitude folds. Rocks that do not deform as easily form long-wavelength, low-amplitude folds.

Fault (geology)

In geology, a fault is a planar fracture or discontinuity in a volume of rock, across which there has been significant displacement along the fractures as a result of rock mass movement. Large faults within the Earth's crust result from the action of plate tectonic forces, with the largest forming the boundaries between the plates, such as subduction zones or transform faults. Energy release associated with rapid movement on active faults is the cause of most earthquakes.

A fault trace or fault line is the intersection of a fault with the ground surface; also, the line commonly plotted on geologic maps to represent a fault.

Since faults do not usually consist of a single, clean fracture, geologists use the term fault zone when referring to the zone of complex deformation associated with the fault plane.

The two sides of a non-vertical fault are known as the hanging wall and footwall. By definition, the hanging wall occurs above the fault plane and the footwall occurs below the fault. This terminology comes from mining: when working a tabular ore body, the miner stood with the footwall under his feet and with the hanging wall hanging above him.

MECHANISMS OF FAULTING

Because of friction and the rigidity of the rock, they cannot glide or flow past each other easily and occasionally all movement is stopped. When this happens, stress builds up in rocks and when it reaches a level that exceeds the strain threshold, the accumulated potential energy is dissipated by the release of strain, which is focused into a plane along which relative motion is accommodated —the fault. Strain is both accumulative and instantaneous depending on the rheology of the rock; the ductile lower crust and mantle accumulates deformation gradually via shearing, whereas the brittle upper crust reacts by fracture - instantaneous stress release - to cause motion along the fault. A fault in ductile rocks can also release instantaneously when the strain rate is too great. The energy released by instantaneous strain release causes earthquakes, a common phenomenon along transform boundaries.

Slip, heave, throw

Slip is defined as the relative movement of geological features present on either side of a fault plane, and is a displacement vector. A fault's sense of slip is defined as the relative motion of the rock on each side of the fault with respect to the other side. In measuring the horizontal or vertical separation, the throw of the fault is the vertical component of the dip separation and the heave of the fault is the horizontal component, as in "throw up and heave out".

The vector of slip can be qualitatively assessed by studying the drag folding of strata on either side of the fault; the direction and magnitude of heave and throw can be measured only by finding common intersection points on either side of the fault (called a piercing point). In practice, it is usually only possible to find the slip direction of faults, and an approximation of the heave and throw vector.

Fault types

Based on direction of slip, faults can be generally categorized as:

- Strike-slip, where the offset is predominately horizontal, parallel to the fault trace.
- Dip-slip, offset is predominately vertical and/or perpendicular to the fault trace.
- Oblique-slip, combining significant strike and dip slip.

Strike-slip faults

The fault surface is usually near vertical and the footwall moves either left or right or laterally with very little vertical motion. Strike-slip faults with left-lateral motion are also known as sinistral faults. Those with right-lateral motion are also known as dextral faults. Each is defined by the direction of movement of the ground on the opposite side of the fault from an observer.

A special class of strike-slip faults is the transform fault, where such faults form a plate boundary. These are found related to offsets in spreading centers, such as mid-ocean ridges, and less commonly within continental lithosphere, such as the San Andreas Fault in California, or the Alpine Fault, New Zealand. Transform faults are also referred to as conservative plate boundaries, as lithosphere is neither created nor destroyed.

Dip-slip faults

Dip-slip faults can occur either as "reverse" or as "normal" faults. A normal fault occurs when the crust is extended. Alternatively such a fault can be called an extensional fault. The hanging wall moves downward, relative to the footwall. A downthrown block between two normal faults dipping towards each other is called a graben. An upthrown block between two normal faults dipping away from each other is called a horst. Low-angle normal faults with regional tectonic significance may be designated detachment faults.

A reverse fault is the opposite of a normal fault—the hanging wall moves up relative to the footwall. Reverse faults indicate compressive shortening of the crust. The dip of a reverse fault is relatively steep, greater than 45°.

A thrust fault has the same sense of motion as a reverse fault, but with the dip of the fault plane at less than 45°. Thrust faults typically form ramps, flats and fault-bend (hanging wall and foot wall) folds. Thrust faults form nappes and klippen in the large thrust belts. Subduction zones are a special class of thrusts that form the largest faults on Earth and give rise to the largest earthquakes.

The fault plane is the plane that represents the fracture surface of a fault. Flat segments of thrust fault planes are known as flats, and inclined sections of the thrust are known as ramps. Typically, thrust faults move within formations by forming flats, and climb up section with ramps.

Fault-bend folds are formed by movement of the hanging wall over a non-planar fault surface and are found associated with both extensional and thrust faults.

Faults may be reactivated at a later time with the movement in the opposite direction to the original movement (fault inversion). A normal fault may therefore become a reverse fault and vice versa.

Oblique-slip faults

A fault which has a component of dip-slip and a component of strike-slip is termed an oblique-slip fault. Nearly all faults will have some component of both dip-slip and strike-slip, so defining a fault as oblique requires both dip and strike components to be measurable and significant. Some oblique faults occur within transtensional and transpressional regimes, others occur where the direction of extension or shortening changes during the deformation but the earlier formed faults remain active.

The hade angle is defined as the complement of the dip angle; it is the angle between the fault plane and a vertical plane that strikes parallel to the fault.

Listric fault

Listric faults are similar to normal faults but the fault plane curves, the dip being steeper near the surface, then shallower with increased depth. The dip may flatten into a sub-horizontal décollement, resulting in horizontal slip on a horizontal plane. The illustration shows slumping of the hanging wall along a listric fault. Where the hanging wall is absent (such as on a cliff) the footwall may slump in a manner that creates multiple listric faults.

Ring fault

Ring faults are faults that occur within collapsed volcanic calderas and the sites of bolide strikes, such as the Chesapeake Bay impact crater. Ring faults may be filled by ring dikes.

Synthetic and antithetic faults

Synthetic and antithetic faults are terms used to describe minor faults associated with a major fault. Synthetic faults dip in the same direction as the major fault while the antithetic faults dip in the opposite direction. Those faults may be accompanied by rollover anticline (e.g. Niger Delta Structural Style).

Fault rock

All faults have a measurable thickness, made up of deformed rock characteristic of the level in the crust where the faulting happened, of the rock types affected by the fault and of the presence and nature of any mineralising fluids. Fault rocks are classified by their textures and the implied mechanism of deformation. A fault that passes through different levels of the lithosphere will have many different types of fault rock developed along its surface. Continued dip-slip displacement tends to juxtapose fault rocks characteristic of different crustal levels, with varying degrees of overprinting. This effect is particularly clear in the case of detachment faults and major thrust faults.

The main types of fault rock include:

⊙ Cataclasite - a fault rock which is cohesive with a poorly developed or absent planar fabric, or which is incohesive, characterised by generally angular clasts and rock fragments in a finer-grained matrix of similar composition.

- Tectonic or Fault breccia - a medium- to coarse-grained cataclasite containing >30% visible fragments.

- Fault gouge - an incohesive, clay-rich fine- to ultrafine-grained cataclasite, which may possess a planar fabric and containing <30% visible fragments. Rock clasts may be present

- Clay smear - clay-rich fault gouge formed in sedimentary sequences containing clay-rich layers which are strongly deformed and sheared into the fault gouge.

◉ Mylonite - a fault rock which is cohesive and characterized by a well-developed planar fabric resulting from tectonic reduction of grain size, and commonly containing rounded porphyroclasts and rock fragments of similar composition to minerals in the matrix

◉ Pseudotachylite - ultrafine-grained vitreous-looking material, usually black and flinty in appearance, occurring as thin planar veins, injection veins or as a matrix to pseudoconglomerates or breccias, which infills dilation fractures in the host rock.

Impacts on structures and people

In geotechnical engineering a fault often forms a discontinuity that may have a large influence on the mechanical behavior (strength, deformation, etc.) of soil and rock masses in, for example, tunnel, foundation, or slope construction.

The level of a fault's activity can be critical for (1) locating buildings, tanks, and pipelines and (2) assessing the seismic shaking and tsunami hazard to infrastructure and people in the vicinity. In California, for example, new building construction has been prohibited directly on or near faults that have moved within the Holocene Epoch (the last 11,000 years). Also, faults that have shown movement during the Holocene plus Pleistocene Epochs (the last 2.6 million years) may receive consideration, especially for critical structures such as power plants, dams, hospitals, and schools. Geologists assess a fault's age by studying soil features seen in shallow excavations and geomorphology seen in aerial photographs. Subsurface clues include shears and their relationships to carbonate nodules, translocated clay, and iron oxide mineralization, in the case of older soil, and lack of such signs in the case of younger soil. Radiocarbon dating of organic material buried next to or over a fault shear is often critical in distinguishing active from inactive faults. From such relationships, paleoseismologists can estimate the sizes of past earthquakes over the past several hundred years, and develop rough projections of future fault activity.

DISCONTINUITIES 25

A discontinuity in geotechnical engineering (in geotechnical literature often denoted by joint) is a plane or surface that marks a change in physical or chemical characteristics in a soil or rock mass. A discontinuity can be, for example, a bedding, schistosity, foliation, joint, cleavage, fracture, fissure, crack, or fault plane. A division is made between mechanical and integral discontinuities. Discontinuities may occur multiple times with broadly the same mechanical characteristics in a discontinuity set, or may be a single discontinuity. A discontinuity makes a soil or rock mass anisotropic.

MECHANICAL DISCONTINUITY

A mechanical discontinuity is a plane of physical weakness where the tensile strength perpendicular to the discontinuity or the shear strength along the discontinuity is lower than that of the surrounding soil or rock material.

Integral discontinuity

An integral discontinuity is a discontinuity that is as strong as the surrounding soil or rock material. Integral discontinuities can change into mechanical discontinuities due to physical or chemical processes (e.g. weathering) that change the mechanical characteristics of the discontinuity.

Discontinuity set or family

Various geological processes create discontinuities at a broadly regular spacing. For example, bedding planes are the result of a repeated sedimentation cycle with a change of sedimentation material or change in structure and texture of the sediment at regular intervals, folding creates joints at regular separations to allow for shrinkage or expansion of the rock material, etc. Normally discontinuities with the same origin have broadly the same characteristics in terms of shear strength, spacing between discontinuities, roughness, infill, etc. The orientations of discontinuities with the same origin are related to the process that has created them and to the geological history of the rock mass. A discontinuity set or family denotes a series of discontinuities for which the geological origin (history, etc.), the orientation, spacing, and the mechanical characteristics (shear strength, roughness, infill material, etc.) are broadly the same.

Single discontinuity

A discontinuity may exist as a single feature (e.g. fault, isolated joint or fracture) and in some circumstances, a discontinuity is treated as a single discontinuity although it belongs to a discontinuity set, in particular if the spacing is very wide compared to the size of the engineering application or to the size of the geotechnical unit.

Discontinuity characterization

Various international standards exist to describe and characterize discontinuities in geomechanical terms, such as ISO 14689-1:2003 and ISRM.

GEOLOGICAL DISCONTINUITIES

The stability of rock slopes is significantly influenced by the structural discontinuity in the rock in which the slope is excavated. A discontinuity is a plane or surface that marks a change in physical or chemical characteristics in a soil or rock mass. A discontinuity can be in the form of a bedding plane, schistosity, foliation, joint, cleavage, fracture, fissure, crack, or fault plane. This discontinuity controls the type of failure which may occur in a rock slope. The properties of discontinuities such as orientation, persistence, roughness and infilling are play important role in the stability of jointed rock slope. Discontinuities may occur multiple times with broadly the same mechanical characteristics in a discontinuity set, or may be a single discontinuity. It makes a soil or rock mass anisotropic.

The orientation of a major geological discontinuity relative to an engineering structure also controls the possibility of unstable conditions and mode of failure. The mutual orientation of discontinuities determines the shape of the individual blocks. Orientation of a discontinuity can be defined by its dip (maximum inclination to the horizontal) and dip direction (direction of the horizontal trace of the line of dip, measured clockwise from north). The strike is at right angles to the dip direction, and the relationship between the strike and the dip direction is illustrated in the possibility if plane failure at lower value of dip angle with respect of slope angle however, as the dip angle of discontinuity increase and become sub parallel to the slope angle the slope become relatively stable However further increase in dip angle in discontinuity make is liable to undergo toppling failure.

A Jointed rock exhibits a higher permeability and, reduced shear strength along the planes of discontinuity apart from increased deformability and negligible tensile strength in directions normal to those planes. The degree of fracturing of a rock mass is controlled by the number of joint in a given direction. A rock mass containing more joints is also considered as more fractured. The spacing of adjacent joints largely controls the size of individual blocks controlling the mode of failure. A close spacing of joints gives low cohesion of rock mass and responsible for circular or even flow failure. It also influences the rock mass permeability.

Persistence of discontinuities defines, together with spacing, the size of blocks that can slide from the face.

Furthermore, a small area of intact rock between low persis-tence discontinuities can have a positive influence on stability because the strength of the rock will often be much higher than the shear stress acting in the slope.

Groundwater

Groundwater is the water located beneath Earth's surface in soil pore spaces and in the fractures of rock formations. A unit of rock or an unconsolidated deposit is called an aquifer when it can yield a usable quantity of water. The depth at which soil pore spaces or fractures and voids in rock become completely saturated with water is called the water table. Groundwater is recharged from, and eventually flows to, the surface naturally; natural discharge often occurs at springs and seeps, and can form oases or wetlands. Groundwater is also often withdrawn for agricultural, municipal, and industrial use by constructing and operating extraction wells. The study of the distribution and movement of groundwater is hydrogeology, also called groundwater hydrology.

Typically, groundwater is thought of as liquid water flowing through shallow aquifers, but, in the technical sense, it can also include soil moisture, permafrost (frozen soil), immobile water in very low permeability bedrock, and deep geothermal or oil formation water. Groundwater is hypothesized to provide lubrication that can possibly influence the movement of faults. It is likely that much of Earth's subsurface contains some water, which may be mixed with other fluids in some instances. Groundwater may not be confined only to Earth. The formation of some of the landforms observed on Mars may have been influenced by groundwater. There is also evidence that liquid water may also exist in the subsurface of Jupiter's moon Europa.

Aquifers

An aquifer is a layer of porous substrate that contains and transmits groundwater. When water can flow directly between the surface and the saturated zone of an aquifer, the aquifer is unconfined. The deeper parts of unconfined aquifers are usually more saturated since gravity causes water to flow downward.

The upper level of this saturated layer of an unconfined aquifer is called the water table or phreatic surface. Below the water table, where in general all pore spaces are saturated with water, is the phreatic zone.

Substrate with low porosity that permits limited transmission of groundwater is known as an aquitard. An aquiclude is a substrate with porosity that is so low it is virtually impermeable to groundwater.

A confined aquifer is an aquifer that is overlain by a relatively impermeable layer of rock or substrate such as an aquiclude or aquitard. If a confined aquifer follows a downward grade from its recharge zone, groundwater can become pressurized as it flows. This can create artesian wells that flow freely without the need of a pump and rise to a higher elevation than the static water table at the above, unconfined, aquifer.

The characteristics of aquifers vary with the geology and structure of the substrate and topography in which they occur. In general, the more productive aquifers occur in sedimentary geologic formations. By comparison, weathered and fractured crystalline rocks yield smaller quantities of groundwater in many environments. Unconsolidated to poorly cemented alluvial materials that have accumulated as valley-filling sediments in major river valleys and geologically subsiding structural basins are included among the most productive sources of groundwater.

The high specific heat capacity of water and the insulating effect of soil and rock can mitigate the effects of climate and maintain groundwater at a relatively steady temperature. In some places where groundwater temperatures are maintained by this effect at about 10 °C (50 °F), groundwater can be used for controlling the temperature inside structures at the surface. For example, during hot weather relatively cool groundwater can be pumped through radiators in a home and then returned to the ground in another well. During cold seasons, because it is relatively warm, the water can be used in the same way as a source of heat for heat pumps that is much more efficient than using air.

Groundwater is often cheaper, more convenient and less vulnerable to pollution than surface water. Therefore, it is commonly used for public water supplies. Groundwater provides the largest source of usable water storage in the United States. Underground reservoirs contain far more water than the capacity of all surface reservoirs and lakes, including the Great Lakes. Many municipal water supplies are derived solely from groundwater.

The volume of groundwater in an aquifer can be estimated by measuring water levels in local wells and by examining geologic records from well-drilling to determine the extent, depth and thickness of water-bearing sediments and rocks. Before an investment is made in production wells, test wells may be drilled to measure the depths at which water is encountered and collect samples of soils, rock and water for laboratory analyses. Pumping tests can be performed in test wells to determine flow characteristics of the aquifer.

Polluted ground water is less visible, but more difficult to clean up, than pollution in rivers and lakes. Ground water pollution most often results from improper disposal of wastes on land. Major sources include industrial and household chemicals and garbage landfills, industrial waste lagoons, tailings and process wastewater from mines, oil field brine pits, leaking underground oil storage tanks and pipelines, sewage sludge and septic systems. Polluted groundwater is mapped by sampling soils and groundwater near suspected or known sources of pollution, to determine the extent of the pollution, and to aid in the design of groundwater reme-diation systems. Preventing groundwater pollution near potential sources such as landfills requires lining the bottom of a landfill with watertight materials, collecting any leachate with drains, and keeping rainwater off any potential contaminants, along with regular monitoring of nearby groundwater to verify that contaminants have not leaked into the groundwater.

The danger of pollution of municipal supplies is minimized by locating wells in areas of deep ground water and impermeable soils, and careful testing and monitoring of the aquifer and nearby potential pollution sources.

Water cycle

Groundwater makes up about twenty percent of the world's fresh water supply, which is about 0.61% of the entire world's water, including oceans and permanent ice. Global groundwater storage is roughly equal to the total amount of freshwater stored in the snow and ice pack, including the north and south poles. This makes it an important resource that can act as a natural storage that can buffer against shortages of surface water, as in during times of drought. Groundwater is naturally replenished by surface water from precipitation, streams, and rivers when this recharge reaches the water table.

Groundwater can be a long-term 'reservoir' of the natural water cycle (with residence times from days to millennia), as opposed to short-term water reservoirs like the atmosphere and fresh surface water (which have residence times from minutes to years). The figure shows how deep groundwater (which is quite distant from the surface recharge) can take a very long time to complete its natural cycle.

The Great Artesian Basin in central and eastern Australia is one of the largest confined aquifer systems in the world, extending for almost 2 million km. By analysing the trace elements in water sourced from deep underground, hydrogeologists have been able to determine that water extracted from these aquifers can be more than 1 million years old.

By comparing the age of groundwater obtained from different parts of the Great Artesian Basin, hydrogeologists have found it increases in age across the basin. Where water recharges the aquifers along the Eastern Divide, ages are young. As groundwater flows westward across the continent, it increases in age, with the oldest groundwater occurring in the western parts. This means that in order to have travelled almost 1000 km from the source of recharge in 1 million years, the groundwater flowing through the Great Artesian Basin travels at an average rate of about 1 metre per year.

Recent research has demonstrated that evaporation of groundwater can play a significant role in the local water cycle, especially in arid regions. Scientists in Saudi Arabia have proposed plans to recapture and recycle this evaporative moisture for crop irrigation. In the opposite photo, a 50-centimeter-square reflective carpet, made of small adjacent plastic cones, was placed in a plant-free dry desert area for five months, without rain or irrigation. It managed to capture and condense enough ground vapor to bring to life naturally buried seeds underneath it, with a green area of about 10% of the carpet area. It is expected that, if seeds were put down before placing this carpet, a much wider area would become green.

ISSUES

Overview

Certain problems have beset the use of groundwater around the world. Just as river waters have been over-used and polluted in many parts of the world, so too have aquifers. The big difference is that aquifers are out of sight. The other major problem is that water management agencies, when calculating the "sustainable yield" of aquifer and river water, have often counted the same water twice, once in the aquifer, and once in its connected river. This problem, although understood for centuries, has persisted, partly through inertia within government agencies. In Australia, for example, prior to the statutory reforms initiated by the Council of Australian Governments water reform framework in the 1990s, many Australian states managed groundwater and surface water through separate government agencies, an approach beset by rivalry and poor communication.

In general, the time lags inherent in the dynamic response of groundwater to development have been ignored by water management agencies, decades after scientific understanding of the issue was consolidated. In brief, the effects of groundwater overdraft (although undeniably real) may take decades or centuries to manifest themselves. In a classic study in 1982, Bredehoeft and colleagues modeled a situation where groundwater extraction in an intermontane basin withdrew the entire annual recharge, leaving 'nothing' for the natural groundwater-dependent vegetation community. Even when the borefield was situated close to the vegetation, 30% of the original vegetation demand could still be met by the lag inherent in the system after 100 years. By year 500, this had reduced to 0%, signalling complete death of the groundwater-dependent vegetation. The science has been available to make these calculations for decades; however, in general water management agencies have ignored effects that will appear outside the rough timeframe of political elections (3 to 5 years). Marios Sophocleous argued strongly that management agencies must define and use appropriate timeframes in groundwater planning. This will mean calculating groundwater withdrawal permits based on predicted effects decades, sometimes centuries in the future.

As water moves through the landscape, it collects soluble salts, mainly sodium chloride. Where such water enters the atmosphere through evapotranspiration, these salts are left behind. In irrigation districts, poor drainage of soils and surface aquifers can result in water tables' coming to the surface in low-lying areas. Major land degradation problems of soil salinity and waterlogging result, combined with increasing levels of salt in surface waters. As a consequence, major damage has occurred to local economies and environments.

Four important effects are worthy of brief mention. First, flood mitigation schemes, intended to protect infrastructure built on floodplains, have had the unintended consequence of reducing aquifer recharge associated with natural flooding. Second, prolonged depletion of groundwater in extensive aquifers can result in land subsidence, with associated infrastructure damage – as well as, third, saline intrusion. Fourth, draining acid sulphate soils, often found in low-lying coastal plains, can result in acidification and pollution of formerly freshwater and estuarine streams.

Another cause for concern is that groundwater drawdown from over-allocated aquifers has the potential to cause severe damage to both terrestrial and aquatic ecosystems – in some cases very conspicuously but in others quite imperceptibly because of the extended period over which the damage occurs.

Overdraft

Groundwater is a highly useful and often abundant resource. However, over-use, or overdraft, can cause major problems to human users and to the environment. The most evident problem (as far as human groundwater use is concerned) is a lowering of the water table beyond the reach of existing wells. As a consequence, wells must be drilled deeper to reach the groundwater; in some places (e.g., California, Texas, and India) the water table has dropped hundreds of feet because of extensive well pumping. In the Punjab region of India, for example, groundwater levels have dropped 10 meters since 1979, and the rate of depletion is accelerating. A lowered water table may, in turn, cause other problems such as groundwater-related subsidence and saltwater intrusion. Groundwater is also ecologically important. The importance of groundwater to ecosystems is often overlooked, even by freshwater biologists and ecologists. Groundwaters sustain rivers, wetlands, and lakes, as well as subterranean ecosystems within karst or alluvial aquifers.

Not all ecosystems need groundwater, of course. Some terrestrial ecosystems – for example, those of the open deserts and similar arid environments – exist on irregular rainfall and the moisture it delivers to the soil, supplemented by moisture in the air. While there are other terrestrial ecosystems in more hospitable environments where groundwater plays no central role, groundwater is in fact fundamental to many of the world's major ecosystems. Water flows between groundwaters and surface waters. Most rivers, lakes, and wetlands are fed by, and (at other places or times) feed groundwater, to varying degrees. Groundwater feeds soil moisture through percolation, and many terrestrial vegetation communities depend directly on either groundwater or the percolated soil moisture above the aquifer for at least part of each year. Hyporheic zones (the mixing zone of streamwater and groundwater) and riparian zones are examples of ecotones largely or totally dependent on groundwater.

Subsidence

Subsidence occurs when too much water is pumped out from underground, deflating the space below the above-surface, and thus causing the ground to collapse. The result can look like craters on plots of land. This occurs because, in its natural equilibrium state, the hydraulic pressure of groundwater in the pore spaces of the aquifer and the aquitard supports some of the weight of the overlying sediments. When groundwater is removed from aquifers by excessive pumping, pore pressures in the aquifer drop and compression of the aquifer may occur. This compression may be partially recoverable if pressures rebound, but much of it is not. When the aquifer gets compressed, it may cause land subsidence, a drop in the ground surface.

The city of New Orleans, Louisiana is actually below sea level today, and its subsidence is partly caused by removal of groundwater from the various aquifer/aquitard systems beneath it. In the first half of the 20th century, the city of San Jose, California dropped 13 feet from land subsidence caused by overpumping; this subsidence has been halted with improved groundwater management.

Seawater intrusion

In general, in very humid or undeveloped regions, the shape of the water table mimics the slope of the surface. The recharge zone of an aquifer near the seacoast is likely to be inland, often at considerable distance. In these coastal areas, a lowered water table may induce sea water to reverse the flow toward the land. Sea water moving inland is called a saltwater intrusion. In alternative fashion, salt from mineral beds may leach into the groundwater of its own accord.

Pollution

Groundwater pollution, from pollutants released to the ground that can work their way down into groundwater, can create a contaminant plume within an aquifer.

Pollution can occur from landfills, naturally occurring arsenic, on-site sanitation systems or other point sources, such as petrol stations or leaking sewers.

Movement of water and dispersion within the aquifer spreads the pollutant over a wider area, its advancing boundary often called a plume edge, which can then intersect with groundwater wells or daylight into surface water such as seeps and springs, making the water supplies unsafe for humans and wildlife. Different mechanism have influence on the transport of pollutants, e.g. diffusion, adsorption, precipitation, decay, in the groundwater. The interaction of groundwater contamination with surface waters is analyzed by use of hydrology transport models.

GEOMORPHOLOGICAL PROCESSES 26

Geomorphology is the scientific study of the origin and evolution of topographic and bathymetric features created by physical or chemical processes operating at or near the earth's surface. Geomorphologists seek to understand why landscapes look the way they do, to understand landform history and dynamics and to predict changes through a combination of field observations, physical experiments and numerical modeling. Geomorphology is practiced within physical geography, geology, geodesy, engineering geology, archaeology and geotechnical engineering. This broad base of interests contributes to many research styles and interests within the field.

OVERVIEW

The surface of the earth is modified by a combination of surface processes that sculpt landscapes, and geologic processes that cause tectonic uplift and subsidence, and shape the coastal geography. Surface processes comprise the action of water, wind, ice, fire, and living things on the surface of the earth, along with chemical reactions that form soils and alter material properties, the stability and rate of change of topography under the force of gravity, and other factors, such as (in the very recent past) human alteration of the landscape. Many of these factors are strongly mediated by climate. Geologic processes include the uplift of mountain ranges, the growth of volcanoes, isostatic changes in land surface elevation (sometimes in response to surface processes), and the formation of deep sedimentary basins where the surface of the earth drops and is filled with material eroded from other parts of the landscape. The earth surface and its topography therefore are an intersection of climatic, hydrologic, and biologic action with geologic processes, or alternatively stated, the intersection of the earth's lithosphere with its hydrosphere, atmosphere, and biosphere.

The broad-scale topographies of the earth illustrate this intersection of surface and subsurface action. Mountain belts are uplifted due to geologic processes. Denudation of these high uplifted regions produces sediment that is transported and deposited elsewhere within the landscape or off the coast. On progressively smaller scales, similar ideas apply, where individual landforms evolve in response to the balance of additive processes (uplift and deposition) and subtractive processes (subsidence and erosion). Often, these processes directly affect each other: ice sheets, water, and sediment are all loads that change topography through flexural isostasy. Topography can modify the local climate, for example through orographic precipitation, which in turn modifies the topography by changing the hydrologic regime in which it evolves. Many geomorphologists are particularly interested in the potential for feedbacks between climate and tectonics, mediated by geomorphic processes.

In addition to these broad-scale questions, geomorphologists address issues that are more specific and/or more local. Glacial geomorphologists investigate glacial deposits such as moraines, eskers, and proglacial lakes, as well as glacial erosional features, to build chronologies of both small glaciers and large ice sheets and understand their motions and effects upon the landscape. Fluvial geomorphologists focus on rivers, how they transport sediment, migrate across the landscape, cut into bedrock, respond to environmental and tectonic changes, and interact with humans. Soils geomorphologists investigate soil profiles and chemistry to learn about the history of a particular landscape and understand how climate, biota, and rock interact. Other geomorphologists study how hillslopes form and change. Still others investigate the relationships between ecology and geomorphology. Because geomorphology is defined to comprise everything related to the surface of the earth and its modification, it is a broad field with many facets.

Geomorphologists use a wide range of techniques in their work. These may include fieldwork and field data collection, the interpretation of remotely sensed data, geochemical analyses, and the numerical modelling of the physics of landscapes. Geomor-phologists may rely on geochronology, using dating methods to measure the rate of changes to the surface. Terrain measurement techniques are vital to quantitatively describe the form of the earth's surface, and include differential GPS, remotely sensed digital terrain models and laser scanning, to quantify, study, and to generate illustrations and maps.

Practical applications of geomorphology include hazard assessment (such as landslide prediction and mitigation), river control and stream restoration, and coastal protection. Planetary geomorphology studies landforms on other terrestrial planets such as Mars. Indications of effects of wind, fluvial, glacial, mass wasting, meteor impact, tectonics and volcanic processes are studied. This effort not only helps better understand the geologic and atmospheric history of those planets but also extends geomor-phological study of the earth. Planetary geomorphologists often use earth analogues to aid in their study of surfaces of other planets.

History

Other than some notable exceptions in antiquity, geomor-phology is a relatively young science, growing along with interest in other aspects of the earth sciences in the mid-19th century. This section provides a very brief outline of some of the major figures and events in its development.

Ancient geomorphology

The first theory of geomorphology was arguably devised by the polymath Chinese scientist and statesman Shen Kuo (1031-1095 AD). This was based on his observation of marine fossil shells in a geological stratum of a mountain hundreds of miles from the Pacific Ocean. Noticing bivalve shells running in a horizontal span along the cut section of a cliffside, he theorized that the cliff was once the pre-historic location of a seashore that had shifted hundreds of miles over the centuries. He inferred that the land was reshaped and formed by soil erosion of the mountains and by deposition of silt, after observing strange natural erosions of the Taihang Mountains and the Yandang Mountain near Wenzhou. Furthermore, he promoted the theory of gradual climate change over centuries of time once ancient petrified bamboos were found to be preserved underground in the dry, northern climate zone of Yanzhou, which is now modern day Yan'an, Shaanxi province.

Early modern geomorphology

The term geomorphology seems to have been first used by Laumann in an 1858 work written in German. Keith Tinkler has suggested that the word came into general use in English, German and French after John Wesley Powell and W. J. McGee used it during the International Geological Conference of 1891.

An early popular geomorphic model was the geographical cycle or cycle of erosion model of broad-scale landscape evolution developed by William Morris Davis between 1884 and 1899. It was an elaboration of the uniformitarianism theory that had first been proposed by James Hutton (1726–1797). With regard to valley forms, for example, uniformitarianism posited a sequence in which a river runs through a flat terrain, gradually carving an increasingly deep valley, until the side valleys eventually erode, flattening the terrain again, though at a lower elevation. It was thought that tectonic uplift could then start the cycle over. In the decades following Davis's development of this idea, many of those studying geomorphology sought to fit their findings into this framework, known today as "Davisian". Davis's ideas are of historical importance, but have been largely superseded today, mainly due to their lack of predictive power and qualitative nature.

In the 1920s, Walther Penck developed an alternative model to Davis's. Penck thought that landform evolution was better described as an alternation between ongoing processes of uplift and denudation, as opposed to Davis's model of a single uplift followed by decay. He also emphasised that in many landscapes slope evolution occurs by backwearing of rocks, not by Davisian-style surface lowering, and his science tended to emphasise surface process over understanding in detail the surface history of a given locality. Penck was German, and during his lifetime his ideas were at times rejected vigorously by the English-speaking geomor-phology community. His early death, Davis' dislike for his work, and his at-times-confusing writing style likely all contributed to this rejection.

Both Davis and Penck were trying to place the study of the evolution of the earth's surface on a more generalized, globally relevant footing than it had been previously. In the early 19th century, authors - especially in Europe - had tended to attribute the form of landscapes to local climate, and in particular to the specific effects of glaciation and periglacial processes. In contrast, both Davis and Penck were seeking to emphasize the importance of evolution of landscapes through time and the generality of the earth's surface processes across different landscapes under different conditions.

During the early 1900s, the study of regional-scale geomor-phology was termed "physiography". Physiography later was considered to be a contraction of "physical" and "geography", and therefore synonymous with physical geography, and the concept became embroiled in controversy surrounding the appropriate concerns of that discipline. Some geomorphologists held to a geological basis for physiography and emphasized a concept of physiographic regions while a conflicting trend among geographers was to equate physiography with "pure morphology," separated from its geological heritage. In the period following World War II, the emergence of process, climatic, and quantitative studies led to a preference by many earth scientists for the term "geomorphology" in order to suggest an analytical approach to landscapes rather than a descriptive one.

Quantitative geomorphology

Geomorphology was started to be put on a solid quantitative footing in the middle of the 20th century. Following the early work of Grove Karl Gilbert around the turn of the 20th century, a group of natural scientists, geologists and hydraulic engineers including Ralph Alger Bagnold, Hans-Albert Einstein, Frank Ahnert, John Hack, Luna Leopold, A. Shields, Thomas Maddock, Arthur Strahler, Stanley Schumm, and Ronald Shreve began to research the form of landscape elements such as rivers and hillslopes by taking systematic, direct, quantitative measurements of aspects of them and investigating the scaling of these measurements. These methods began to allow prediction of the past and future behavior of landscapes from present observations, and were later to develop into the modern trend of a highly quantitative approach to geomorphic problems. Quantitative geomorphology can involve fluid dynamics and solid mechanics, geomorphometry, laboratory studies, field measurements, theoretical work, and full landscape evolution modeling. These approaches are used to understand weathering and the formation of soils, sediment transport, landscape change, and the interactions between climate, tectonics, erosion, and deposition.

Contemporary geomorphology

Today, the field of geomorphology encompasses a very wide range of different approaches and interests. Modern researchers aim to draw out quantitative "laws" that govern earth surface processes, but equally, recognize the uniqueness of each landscape and environment in which these processes operate. Particularly important realizations in contemporary geomorphology include:

- that not all landscapes can be considered as either "stable" or "perturbed", where this perturbed state is a temporary displacement away from some ideal target form. Instead, dynamic changes of the landscape are now seen as an essential part of their nature.

- that many geomorphic systems are best understood in terms of the stochasticity of the processes occurring in them, that is, the probability distributions of event magnitudes and return times. This in turn has indicated the importance of chaotic determinism to landscapes, and that landscape properties are best considered statistically. The same processes in the same landscapes do not always lead to the same end results.

Processes

Geomorphically relevant processes generally fall into (1) the production of regolith by weathering and erosion, (2) the transport of that material, and (3) its eventual deposition. Primary surface processes responsible for most topographic features include wind, waves, chemical dissolution, mass wasting, groundwater movement, surface water flow, glacial action, tectonism, and volcanism. Other more exotic geomorphic processes might include periglacial (freeze-thaw) processes, salt-mediated action, or extraterrestrial impact.

Eolian processes

Eolian processes pertain to the activity of the winds and more specifically, to the winds' ability to shape the surface of the earth. Winds may erode, transport, and deposit materials, and are effective agents in regions with sparse vegetation and a large supply of fine, unconsolidated sediments. Although water and mass flow tend to mobilize more material than wind in most environments, eolian processes are important in arid environments such as deserts.

Biological processes

The interaction of living organisms with landforms, or biogeomorphologic processes, can be of many different forms, and is probably of profound importance for the terrestrial geomorphic system as a whole. Biology can influence very many geomorphic processes, ranging from biogeochemical processes controlling chemical weathering, to the influence of mechanical processes like burrowing and tree throw on soil development, to even controlling global erosion rates through modulation of climate through carbon dioxide balance. Terrestrial landscapes in which the role of biology in mediating surface processes can be definitively excluded are extremely rare, but may hold important information for understanding the geomorphology of other planets, such as Mars.

Fluvial processes

Rivers and streams are not only conduits of water, but also of sediment. The water, as it flows over the channel bed, is able to mobilize sediment and transport it downstream, either as bed load, suspended load or dissolved load. The rate of sediment transport depends on the availability of sediment itself and on the river's discharge. Rivers are also capable of eroding into rock and creating new sediment, both from their own beds and also by coupling to the surrounding hillslopes. In this way, rivers are thought of as setting the base level for large scale landscape evolution in nonglacial environments. Rivers are key links in the connectivity of different landscape elements. As rivers flow across the landscape, they generally increase in size, merging with other rivers. The network of rivers thus formed is a drainage system. These systems take on four general patterns, dendritic, radial, rectangular, and trellis. Dendritic happens to be the most common occurring when the underlying strata is stable (without faulting). Drainage systems have four primary components: drainage basin, alluvial valley, delta plain, and receiving basin. Some geomorphic examples of fluvial landforms are alluvial fans, oxbow lakes, and fluvial terraces.

Glacial processes

Glaciers, while geographically restricted, are effective agents of landscape change. The gradual movement of ice down a valley causes abrasion and plucking of the underlying rock. Abrasion produces fine sediment, termed glacial flour. The debris transported by the glacier, when the glacier recedes, is termed a moraine. Glacial erosion is responsible for U-shaped valleys, as opposed to the V-shaped valleys of fluvial origin.

The way glacial processes interact with other landscape elements, particularly hillslope and fluvial processes, is an important aspect of Plio-Pleistocene landscape evolution and its sedimentary record in many high mountain environments. Environments that have been relatively recently glaciated but are no longer may still show elevated landscape change rates compared to those that have never been glaciated. Nonglacial geomorphic processes which nevertheless have been conditioned by past glaciation are termed paraglacial processes. This concept contrasts with periglacial processes, which are directly driven by formation or melting of ice or frost.

Hillslope processes

Soil, regolith, and rock move downslope under the force of gravity via creep, slides, flows, topples, and falls. Such mass wasting occurs on both terrestrial and submarine slopes, and has been observed on Earth, Mars, Venus, Titan and Iapetus.

Ongoing hillslope processes can change the topology of the hillslope surface, which in turn can change the rates of those processes. Hillslopes that steepen up to certain critical thresholds are capable of shedding extremely large volumes of material very quickly, making hillslope processes an extremely important element of landscapes in tectonically active areas. On the earth, biological processes such as burrowing or tree throw may play important roles in setting the rates of some hillslope processes.

Igneous processes

Both volcanic (eruptive) and plutonic (intrusive) igneous processes can have important impacts on geomorphology. The action of volcanoes tends to rejuvenize landscapes, covering the old land surface with lava and tephra, releasing pyroclastic material and forcing rivers through new paths. The cones built by eruptions also build substantial new topography, which can be acted upon by other surface processes. Plutonic rocks intruding then solidifying at depth can cause both uplift or subsidence of the surface, depending on whether the new material is denser or less dense than the rock it displaces.

Tectonic processes

Tectonic effects on geomorphology can range from scales of millions of years to minutes or less. The effects of tectonics on landscape are heavily dependent on the nature of the underlying bedrock fabric that more less controls what kind of local morphology tectonics can shape. Earthquakes can, in terms of minutes, submerge large areas of land creating new wetlands. Isostatic rebound can account for significant changes over hundreds to thousands of years, and allows erosion of a mountain belt to promote further erosion as mass is removed from the chain and the belt uplifts. Long-term plate tectonic dynamics give rise to orogenic belts, large mountain chains with typical lifetimes of many tens of millions of years, which form focal points for high rates of fluvial and hillslope processes and thus long-term sediment production.

Features of deeper mantle dynamics such as plumes and delamination of the lower lithosphere have also been hypothesised to play important roles in the long term (> million year), large scale (thousands of km) evolution of the earth's topography. Both can promote surface uplift through isostasy as hotter, less dense, mantle rocks displace cooler, denser, mantle rocks at depth in the earth.

Scales in geomorphology

Different geomorphological processes dominate at different spatial and temporal scales. Moreover, scales on which processes occur may determine the reactivity or otherwise of landscapes to changes in driving forces such as climate or tectonics. These ideas are key to the study of geomorphology today.

To help categorize landscape scales some geomorphologists might use the following taxonomy:

- ◉ 1st - Continent, ocean basin, climatic zone (~10,000,000 km)
- ◉ 2nd - Shield, e.g. Baltic Shield, or mountain range (~1,000,000 km)
- ◉ 3rd - Isolated sea, Sahel (~100,000 km)
- ◉ 4th - Massif, e.g. Massif Central or Group of related landforms, e.g., Weald (~10,000 km)
- ◉ 5th - River valley, Cotswolds (~1,000 km)
- ◉ 6th - Individual mountain or volcano, small valleys (~100 km)
- ◉ 7th - Hillslopes, stream channels, estuary (~10 km)
- ◉ 8th - gully, barchannel (~1 km)
- ◉ 9th - Meter-sized features

BIBLIOGRAPHY

Amelin, Y. (2002). Lead Isotopic Ages of Chondrules and Calcium-Aluminum-Rich Inclusions. Science 297 (5587): 1678. doi:10.1126/science.1073950. edit

B.A. van der Pluijm and S. Marshak (2004). Earth Structure - An Introduction to Structural Geology and Tectonics (2nd ed.). New York: W. W. Norton. p. 656. ISBN 0-393-92467-X.

C.W. Passchier and R.A.J. Trouw (1998). Microtectonics. Berlin: Springer. ISBN 3-540-58713-6.

Condomines, M; Tanguy, J; Michaud, V (1995). Magma dynamics at Mt Etna: Constraints from U-Th-Ra-Pb radioactive disequilibria and Sr isotopes in historical lavas. Earth and Planetary Science Letters 132: 25. Bibcode:1995E&PSL.132...25C. doi:10.1016/0012-821X(95)00052-E.

Dahlen, F A (1990). Critical Taper Model of Fold-And-Thrust Belts and Accretionary Wedges. Annual Review of Earth and Planetary Sciences 18: 55. Bibcode:1990AREPS..18...55D. doi:10.1146/annurev.ea.18.050190.000415.

Dalrymple, G. Brent (1994). The age of the earth. Stanford, California: Stanford Univ. Press. ISBN 0-8047-2331-1.

Faure, Gunter (1998). Principles and applications of geochemistry: a comprehensive textbook for geology students. Upper Saddle River, NJ: Prentice-Hall. ISBN 978-0-02-336450-1.

G.H. Davis and S.J. Reynolds (1996). The structural geology of rocks and regions (2nd ed.). Wiley. ISBN 0-471-52621-5.

Henkel, D. J. (1960), Undrained Shear Strength of Anisotropically Consolidated Clays, ASCE Speciality Conference on Shear Strength of Cohesive Soils, University of Colorado, Boulder, Colo., June 13-17: 533–554

Hess, H. H. (November 1, 1962) History Of Ocean Basins, pp. 599–620 in Petrologic studies: a volume in honor of A. F. Buddington. A. E. J. Engel, Harold L. James, and B. F. Leonard (eds.) [New York?]: Geological Society of America.

Heyman, J. (1972), Coulomb's Memoir on Statics, Cambridge University Press, ISBN 978-1-86094-056-9

Hubbard, Bryn and Glasser, Neil (2005). Field techniques in glaciology and glacial geomorphology. Chichester, England: J. Wiley. ISBN 0-470-84426-4.

Jonathan T. Ricketts, M. Kent Loftin, Frederick S. Merritt, ed. (2004). Standard handbook for civil engineers (5 ed.). McGraw Hill.ISBN 0071364730.

Joseph, P. G. (2009), Constitutive Model of Soil Based on a Dynamical Systems Approach (PDF), Journal of Geotechnical and Geoenvironmental Engineering 135 (No. 8): 1155–1158.

Joseph, P. G. (2012), Physical Basis and Validation of a Constitutive Model for Soil Shear Derived from Micro Structural Changes (PDF), International Journal of Geomechanics.

Kious, Jacquelyne; Tilling, Robert I. (February 1996). Understanding Plate Motions. This Dynamic Earth: The Story of Plate Tectonics. Kiger, Martha, Russel, Jane (Online ed.). Reston, Virginia, USA: United States Geological Suvey. ISBN 0-16-048220-8. Retrieved 13 March 2009.

Krumbein, Wolfgang E., ed. (1978). Environmental biogeo-chemistry and geomicrobiology. Ann Arbor, Mich.: Ann Arbor Science Publ. ISBN 0-250-40218-1.

Ladd, C. C.; Foott, R. (1974), New Design Procedure for Stability of Soft Clays, Journal of Geotechnical Engineering 100 (GT7): 763–786

Levin, Harold L. (2010). The earth through time (9th ed.). Hoboken, N.J.: J. Wiley. p. 18. ISBN 978-0-470-38774-0.

Liddell, Henry George; Scott, Robert; A Greek–English Lexicon at the Perseus Project

M. King Hubbert (1972). Structural Geology. Hafner Publishing Company.

McBirney, Alexander R. (2007). Igneous petrology. Boston: Jones and Bartlett Publishers. ISBN 978-0-7637-3448-0.

McDougall, Ian; Harrison, T. Mark (1999). Geochronology and thermochronology by the o&°Ar/©Ar method. New York: Oxford University Press. ISBN 0-19-510920-1.

Morton, A. C. (1985). A new approach to provenance studies: electron microprobe analysis of detrital garnets from Middle Jurassic sandstones of the northern North Sea. Sedimentology 32 (4): 553. Bibcode:1985Sedim..32..553M. doi:10.1111/j.1365-3091.1985.tb00470.x.

Oxford Dictionary of National Biography, 1961-1970 (Roscoe, Kenneth Harry): 894–896 Missing or empty |title= (help)

Patterson, C. (1956). Age of Meteorites and the Earth. Geochimica et Cosmochimica Acta 10: 230–237.

Poulos, S. J. (1971), The Stress-Strain Curve of Soils (PDF), GEI Internal Report

Poulos, S. J. (1981), The Steady State of Deformation, Journal of Geotechnical Engineering 107 (GT5): 553–562

Poulos, S. J. (1989), Jansen, R. B., ed., Liquefaction Related Phenomena (PDF), Advance Dam Engineering for Design (Van Nostrand Reinhold): 292–320

Reijer Hooykaas, Natural Law and Divine Miracle: The Principle of Uniformity in Geology, Biology, and Theology, Leiden: EJ Brill, 1963.

Richard Greenburg (2005). The Ocean Moon: Search for an Alien Biosphere. Springer Praxis Books.

Roscoe, K. H.; Schofield, A. N.; Wroth, C. P. (1958), On the Yielding of Soils, Geotechnique 8: 22–53

Shepherd, T.J.; Rankin, A.H. and Alderton, D.H.M. (1985). A practical guide to fluid inclusion studies. Glasgow: Blackie. ISBN 0-412-00601-4.

Terzaghi, K. (1942), Theoretical Soil Mechanics, New York: Wiley, ISBN 978-0-471-85305-3

USGS Topographic Maps. United States Geological Survey. Retrieved 2009-04-11.

V. Guerriero et al. (2009). Quantifying uncertainties in multi-scale studies of fractured reservoir analogues: Implemented statistical analysis of scan line data from carbonate rocks. Journal of Structural Geology (Elsevier) 32 (9): 1271–1278.

V. Guerriero et al. (2011). Improved statistical multi-scale analysis of fractures in carbonate reservoir analogues. Tectonophysics (Elsevier) 504: 14–24.

W.F. Chen and J.Y. Richard Liew, ed. (2002). The Civil Engineering Handbook. CRC Press. ISBN 978-0-8493-0958-8.

Wegener, A. (1999). Origin of continents and oceans. Courier Corporation. ISBN 0-486-61708-4.

Zheng, Y; Fu, Bin; Gong, Bing; Li, Long (2003). Stable isotope geochemistry of ultrahigh pressure metamorphic rocks from the Dabie–Sulu orogen in China: implications for geodynamics and fluid regime. Earth-Science Reviews 62: 105. Bibcode: 2003ESRv...62..105Z. doi:10.1016/S0012-8252(02)00133-2.

INDEX

www.ingramcontent.com/pod-product-compliance
Lightning Source LLC
Chambersburg PA
CBHW082005190326
41458CB00010B/3074